© 1992-2004 www.nemmar.com

Real Estate Home Inspection Checklist From <u>A</u> to <u>Z</u>

Just a few of the many excellent book reviews our products have received:

Top 50 book reviewers for **Amazon.com** have given our books the highest rating of **5 stars!**

Real Estate Press - "…the Real Estate from A to Z books and videos are the best we've ever seen."

New Home Construction Journal – "…the best selling reference books available for home builders and buyers and cover every topic from A to Z."

Seminar Progress Report - "…top-notch real estate investors, inspectors and appraisers agree the Real Estate from A to Z series is a great value."

Home-Based Business Monthly - "If you're looking to become a knowledgeable home inspector or appraiser, Real Estate from A to Z series is crucial."

Real Estate Investors Journal - "…Real Estate from A to Z series is by far the most in-depth resource for every investor, beginner to expert."

⭐⭐⭐⭐⭐ **Everything you must know to inspect a home.**
Amazon Top 50 Book Reviewer from Florida
Home Inspection Business from A to Z

⭐⭐⭐⭐⭐ **Learn the flaws to look for in a home.**
Amazon Top 50 Book Reviewer from Iowa
Home Inspection Business from A to Z

⭐⭐⭐⭐⭐ **Knocks it out of the box!**
Amazon Top 50 Book Reviewer
Home Inspection Business from A to Z

⭐⭐⭐⭐⭐ **Home Inspector Know-How.**
Reviewer: Author/Reviewer from California
Home Inspection Business from A to Z

⭐⭐⭐⭐⭐ **A thorough exposition of how to appraise real estate.**
Amazon Top 50 Book Reviewer from Florida
Real Estate Appraisal from A to Z

⭐⭐⭐⭐⭐ **Appraisal Know-How.**
Reviewer: Author/Reviewer from California
Real Estate Appraisal from A to Z

⭐⭐⭐⭐⭐ **Appraiser's Dream Book for beginners or experts!**
Reviewer: Author/Reviewer from New York
Real Estate Appraisal from A to Z

⭐⭐⭐⭐⭐ **Highly Recommended book for real estate appraisers.**
Managing Editor: Book Review Club Top Reviewer from California
Real Estate Appraisal from A to Z

Graph Pages

What's In This For You?

Home Inspection Business From A to Z book - Home Inspectors earn **$400** or more for each inspection. Many inspectors are so busy that they do two home inspections per day!

You don't need to have any background in real estate, construction, nor engineering to become a highly paid home inspector. All you need is to obtain the right knowledge and business plan, *(which we'll give you)*. Male or female, and your current age doesn't matter either, since there is *no* manual labor or hard work involved.

Here's just a few of the
benefits you get from this book:

Earn Money in one of the fastest growing businesses in the country.

Be Your Own Boss and work part-time or full-time. *You* set your own hours.

Save Money when you buy, sell, or renovate your own home.

Eliminate Safety Hazards to make your home safe for you and your family.

Real Estate Related knowledge to help people with the *biggest* investment they will ever make - their own home.

What's The Income Potential?

Are you wondering about the income and growth potential of the home inspection business? Then look below because a picture is worth a thousand words. Home inspectors earn **high** incomes every year while working right out of their home. The home inspection industry has been growing by leaps and bounds every year. And remember, this growth has gone on even during some of the worst economic recessions in history! Our *Home Inspection Business From A to Z* book shows you how your home inspection business can grow even during a recession.

Growth Of Home Inspection Industry

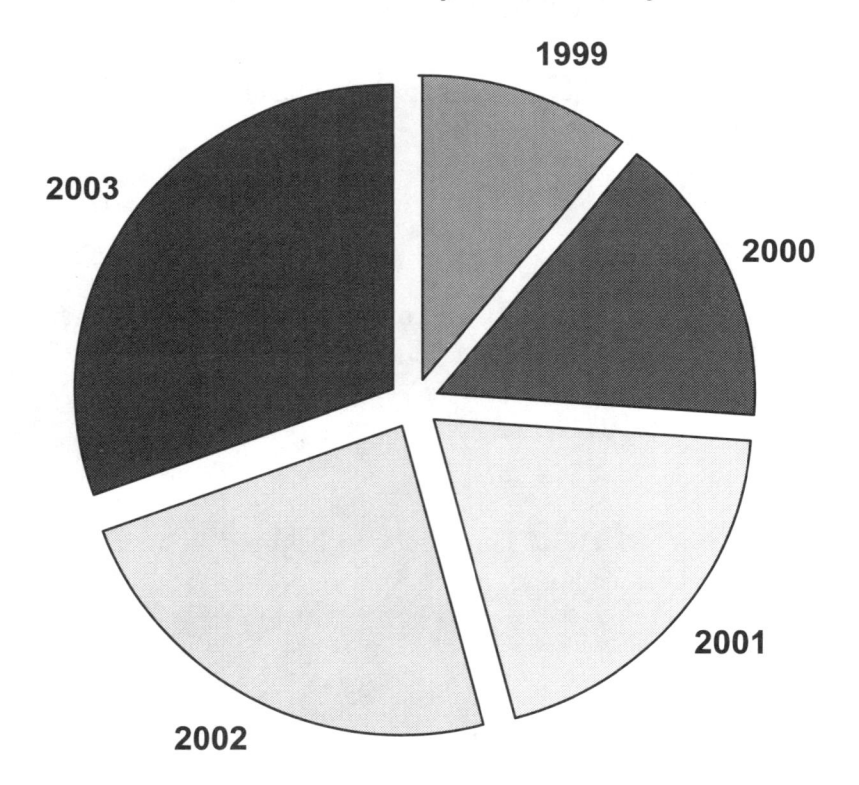

What Houses Need Inspections?

Are you wondering about the need for home inspections? Then look below because this data has been taken from evaluating thousands of homes. The graph shows the probability of finding repair problems in any house, *(including yours!)* This book shows you the reasons why <u>all</u> homes need to be inspected.
Do you really know the true condition of your house?

Average Home Repairs Needed

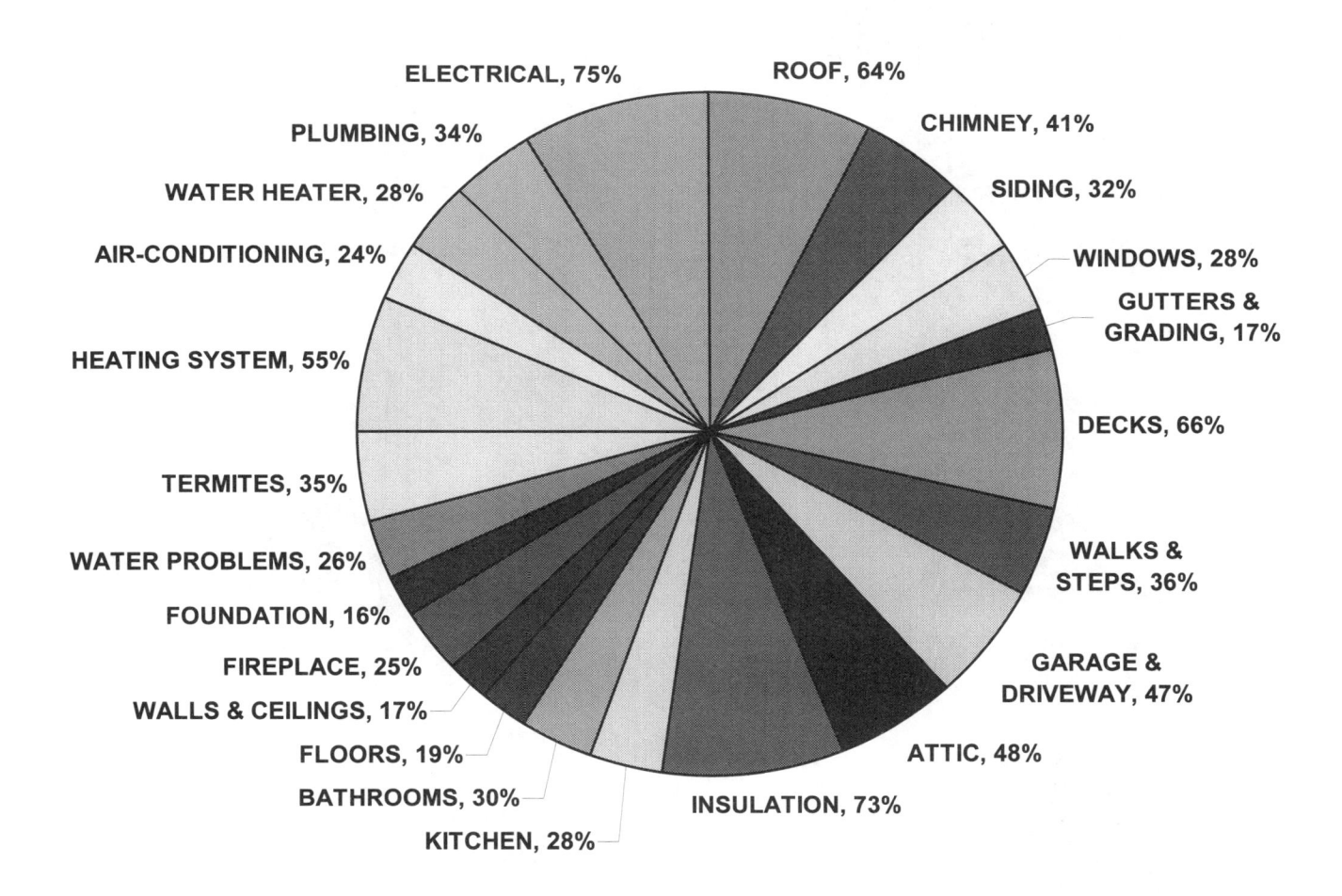

- ELECTRICAL, 75%
- ROOF, 64%
- PLUMBING, 34%
- CHIMNEY, 41%
- WATER HEATER, 28%
- SIDING, 32%
- AIR-CONDITIONING, 24%
- WINDOWS, 28%
- GUTTERS & GRADING, 17%
- HEATING SYSTEM, 55%
- DECKS, 66%
- TERMITES, 35%
- WALKS & STEPS, 36%
- WATER PROBLEMS, 26%
- FOUNDATION, 16%
- GARAGE & DRIVEWAY, 47%
- FIREPLACE, 25%
- WALLS & CEILINGS, 17%
- ATTIC, 48%
- FLOORS, 19%
- INSULATION, 73%
- BATHROOMS, 30%
- KITCHEN, 28%

What's A Buyer's Savings?

So you don't want a career change? That's fine, but would you be interested in saving thousands of dollars when you buy your own home? Then look below. This graph shows some potential repair costs found when buying your home. By identifying any problem conditions *before* you buy, you'll be able to negotiate a lower purchase price.

Average $ Savings For Home Buyers

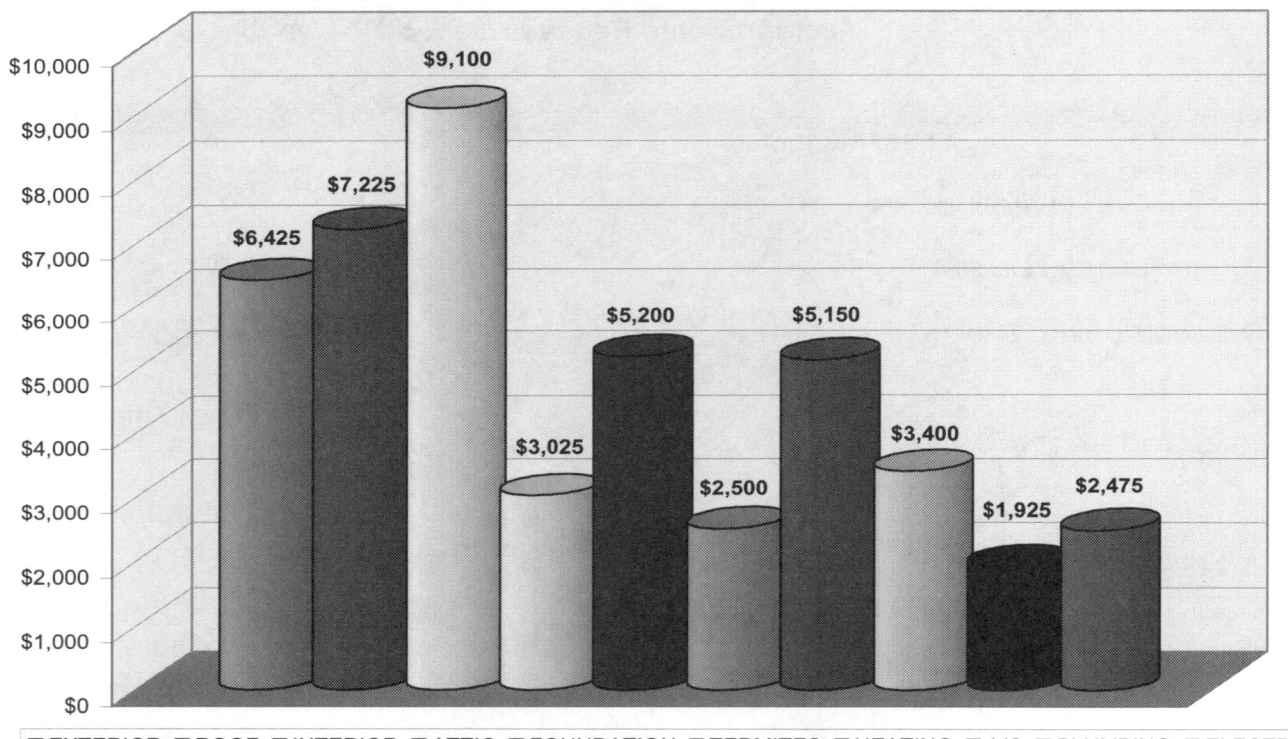

EXTERIOR	$6,425
ROOF	$7,225
INTERIOR	$9,100
ATTIC	$3,025
FOUNDATION	$5,200
TERMITES	$2,500
HEATING	$5,150
A/C	$3,400
PLUMBING	$1,925
ELECTRIC	$2,475

Don't Let Your Dream House...

What's A Seller's Savings?

You're not buying a house at this time? Okay, but would you be interested in saving thousands of dollars when you renovate or sell your own home? Then look down. This graph shows your potential profit when selling your home. By eliminating these typical repairs you can earn at least an additional $1.50 for each $1.00 you invest in repair expenses.

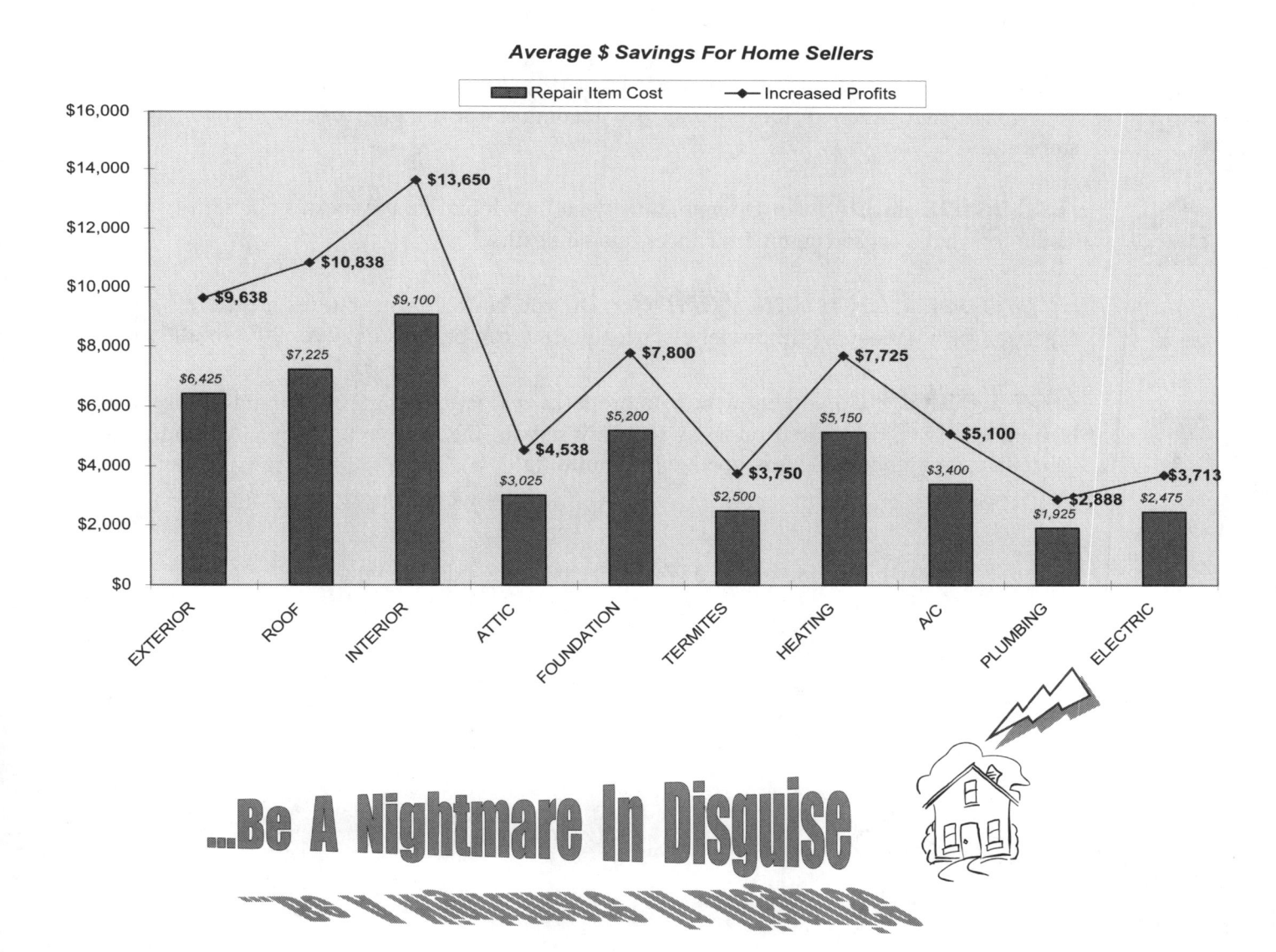

Average $ Savings For Home Sellers

Repair Item Cost | Increased Profits

Category	Repair Item Cost	Increased Profits
EXTERIOR	$6,425	$9,638
ROOF	$7,225	$10,838
INTERIOR	$9,100	$13,650
ATTIC	$3,025	$4,538
FOUNDATION	$5,200	$7,800
TERMITES	$2,500	$3,750
HEATING	$5,150	$7,725
A/C	$3,400	$5,100
PLUMBING	$1,925	$2,888
ELECTRIC	$2,475	$3,713

...Be A Nightmare In Disguise

What Safety Hazards?

So you think there are no safety hazards around your home?
Well, think again.

Lead In Water & Paint - Do you know how to check for lead in your home? Lead poisoning is the <u>number one</u> childhood disease in America!

Radon Gas - Is it in your house? The EPA, *(Environmental Protection Agency)*, has determined that radon is the <u>number two</u> leading cause of lung cancer behind cigarette smoking!

Asbestos & UFFI - Do you know what they look like? Asbestos and UFFI insulations have caused countless cancer related deaths.

Improper Electrical Wiring - Do you know if your outlets meet the *National Electric Code* standards? Improper electrical wiring can be found in over 90% of all homes!

Gas Leaks - Do you know how to properly evaluate your gas meter and supply lines? Natural gas is colorless and odorless before it gets to the utility company. An undetected gas leak can explode and blow up an entire building!

We're not trying to scare you, we're trying to educate you.
We just want to open your eyes to the reality of some of the
safety hazards that can be found in *any* home.

Customer Comments & Recommendations

From the Author

Thank you very much for purchasing my book. I invite you to view our web site at **www.nemmar.com** to see the other real estate products we offer that will save you thousands of dollars when you buy, sell, or renovate a home. You can sign up online to receive our **free** real estate newsletter with articles and updates that will definitely help you profit in real estate. Please email me and let me know what you think of my book after you have time to review it. Customer feedback and recommendations are greatly appreciated and help me to improve all of our products.
Thank you,

Guy Cozzi
Nemmar Real Estate Training
*"Everything You Need To Know About Real Estate
From **A**sbestos to **Z**oning"*

Nemmar Real Estate Training is ranked as the most exclusive real estate appraisal, home inspection and real estate investment training service since 1988. Our real estate books, DVDs, CDs and Videos are rated **number one** in the Real Estate Appraisal, Home Inspection and Home Improvement categories nationwide! Our products have taught thousands of home buyers, sellers, and real estate professionals worldwide. You too can learn everything you need to know about Real Estate - from **A**sbestos to **Z**oning. With this knowledge you will save thousands of dollars when you buy, sell, or renovate your home. You will also learn how to eliminate safety hazards and properly maintain a home. Statistics show an average savings of at least **$4,700.00** per home for customers who have read our books.

Our home inspection, appraisal, and home improvement books have been called the "Bible" of the real estate industry. Written by Guy Cozzi who has decades of experience as a licensed appraiser, home inspector, consultant, and real estate investor. This top selling author has been quoted as a real estate expert by the *New York Times* and many other publications. He has been a guest speaker on real estate investment TV shows and has taught thousands of people how to inspect, appraise and invest in real estate and provides advice to many banks and mortgage lenders.

The real facts other books don't tell you! ***You'll learn everything that your Realtor doesn't want you to know.*** Realtors "sugar coat" the problem conditions in a house in order to close the deal and get paid their sales commission. This is unquestionably the only book of its kind that teaches you how to prevent those pitfalls. You get information that the professionals use to make you an educated consumer enabling you to negotiate a much better price on the purchase, renovation, or sale of your home.

Do you really know the true condition of your house? Don't let your **dream house** be a **nightmare** in disguise! Our *Real Estate From A to Z* books and DVDs will assist you with the biggest investment of your life - your own home! These products were originally designed to train top-notch, professional home inspectors, appraisers, investors, builders, contractors and Realtors and are now available at a price affordable to everyone.

**All homes need to be inspected, appraised and updated for safety hazards,
routine maintenance, and energy efficiency.**

© Copyright 1992-2004 Guy Cozzi, All Rights Reserved. Email: info@nemmar.com
No part of this work shall be reproduced, stored in a retrieval system, or transmitted by any means, electronic, mechanical, photocopying, recording or otherwise without the written permission of Guy Cozzi. This document and accompanying materials are designed to provide authoritative information in regard to the subject matter covered in it. It is sold with the understanding that the author and publisher are not engaged in rendering legal, accounting or other professional opinions. If legal advice or other expert assistance is required, the services of a competent professional should be sought. *(This statement is from a declaration of principles adopted jointly by a committee of publishers and associations).* All **NeMMaR** products and all other brand and product names are trademarks or registered trademarks of their respective holders.

Table of Contents

Real Estate Home Inspection Checklist From A to Z 1
   Graph Pages 3
   From the Author 9

Table of Contents 10

Introduction 13
   Introduction 13
   Purpose Of A Home Inspection 14
   Benefits Of Knowledge Of Home Inspections 14
   Description Of A Home Inspection 15
   Professional Engineer and Registered Architect Issue .. 16
   Tools That Are Helpful 17
   Booking Home Inspection Jobs 18
   Beginning The Home Inspection 21
   Definition Of Terms 22

Sample Inspection Report Pages 23

Subject Property #1 28

Questions To Ask The Home Seller #1 29
   ◊ Age, Zoning, and Permits: 29
   ◊ Interior Inspection: 29
   ◊ Exterior Inspection: 29
   ◊ Operating Systems: 30
   ◊ Termite and Water Problems: 31
   ◊ Septic System: 31
   ◊ Well Water System: 32
   ◊ Swimming Pool: 32

The Operating Systems Inspection #1 33
   Heating System 33
   ◊ Oil Fired Heating Systems: 34
   ◊ Gas Fired Heating Systems: 35
   ◊ Electric Heating Systems: 35
   ◊ Forced Warm Air Systems: 36
   ◊ Heat Pump Heating Systems: 37
   ◊ Forced Hot Water Heating Systems: 38
   ◊ Steam Heating Systems: 39
   Air-Conditioning System 40
   Domestic Water Heater 42
   ◊ Separate Domestic Water Heaters: 42
   ◊ Immersion Coil Water Heaters: 43
   ◊ Oil Fired Water Heaters: 44
   ◊ Gas Fired Water Heaters: 44
   ◊ Electrically Operated Water Heaters: 44
   Plumbing System 45
   Well Water System 46
   Septic System 48
   Electrical System 49
   Additional Comments 52

The Lower Level Inspection #1 53
   Lower Level 53
   Crawl Spaces 55
   Gas Service 56
   Auxiliary Systems 57
   Water Penetration 57
   Additional Comments 59

The Interior Home Inspection #1 61
   Kitchen 61
   Bathrooms 62
   Floors and Stairs 63
   Walls and Ceilings 63
   Windows and Doors 64
   Fireplaces 65
   Attic Inspection 66
   Attic Ventilation 67
   Attic Insulation 68

Asbestos Insulation	69
Radon Gas	70
Additional Comments	71

The Exterior Home Inspection #1 73

Roof	73
Chimney	74
Siding	74
Fascia, Soffits and Eaves	75
Gutters, Downspouts and Leaders	76
Windows, Screens and Storms	76
Entrances, Steps and Porches	77
Walks	77
Patios and Terraces	78
Decks	78
Walls and Fences	79
Drainage and Grading	80
Driveways	80
Garage	81
Other Exterior Structures	81
Swimming Pools	82
Wood Destroying Insects	82
Additional Comments	85

Safety Concerns #1 86

Safety Concerns	86
Additional Comments	87

Home Inspection Conclusion #1 88

Conclusion	88

Home Inspection Photo Pages 89

More Nemmar Products	106

Subject Property #2 108

Questions To Ask The Home Seller #2 109

◊ Age, Zoning, and Permits:	109
◊ Interior Inspection:	109
◊ Exterior Inspection:	109
◊ Operating Systems:	110
◊ Termite and Water Problems:	111
◊ Septic System:	111
◊ Well Water System:	112
◊ Swimming Pool:	112

The Operating Systems Inspection #2 113

Heating System	113
◊ Oil Fired Heating Systems:	114
◊ Gas Fired Heating Systems:	115
◊ Electric Heating Systems:	115
◊ Forced Warm Air Systems:	116
◊ Heat Pump Heating Systems:	117
◊ Forced Hot Water Heating Systems:	118
◊ Steam Heating Systems:	119
Air-Conditioning System	120
Domestic Water Heater	122
◊ Separate Domestic Water Heaters:	122
◊ Immersion Coil Water Heaters:	123
◊ Oil Fired Water Heaters:	124
◊ Gas Fired Water Heaters:	124
◊ Electrically Operated Water Heaters:	124
Plumbing System	125
Well Water System	126
Septic System	128
Electrical System	129
Additional Comments	132

The Lower Level Inspection #2 133

Lower Level	133
Crawl Spaces	135
Gas Service	136
Auxiliary Systems	137
Water Penetration	137
Additional Comments	139

The Interior Home Inspection #2 141

Kitchen	141
Bathrooms	142
Floors and Stairs	143
Walls and Ceilings	143
Windows and Doors	144
Fireplaces	145
Attic Inspection	146

Attic Ventilation 147
Attic Insulation 148
Asbestos Insulation 149
Radon Gas 150
Additional Comments 151

The Exterior Home Inspection #2 .... *153*

Roof 153
Chimney 154
Siding 154
Fascia, Soffits and Eaves 155
Gutters, Downspouts and Leaders 156
Windows, Screens and Storms 156
Entrances, Steps and Porches 157
Walks 157
Patios and Terraces 158
Decks 158
Walls and Fences 159
Drainage and Grading 160
Driveways 160
Garage 161
Other Exterior Structures 161
Swimming Pools 162
Wood Destroying Insects 162
Additional Comments 165

Safety Concerns #2 .... *166*

Safety Concerns 166
Additional Comments 167

Home Inspection Conclusion #2 .... *168*

Conclusion 168
More Nemmar Products 169

Index *172*

Introduction

Introduction

This book is going to cover every aspect of a real estate home inspection checklist from A to Z. This checklist was originally designed to train top-notch, professional real estate home inspectors. However, this book is extremely helpful to anyone involved in real estate. This includes a home buyer, homeowner, home seller, Realtor, appraiser, etc. Refer to the **Benefits Of Knowledge Of Home Inspections** section to see the reasons why **anyone** can benefit by knowing this material. A lot of the information in this educational material goes well beyond what's covered in other books. That's why I say it's "the *real facts* other books don't tell you!"

You may have purchased some of my other books and DVDs. If you have, then you will find that some information contained in this book is similar. This is because there are many important aspects that pertain to both the real estate appraisal and the home inspection businesses, as well as, to real estate investing. I hope you don't feel that some of the information is redundant. It will only benefit you to read it several times to make sure you know the information well enough so it doesn't cost you time and money later. If you haven't purchased some of our other products, then do it now! I'm serious about that because our products are worth much more than the price we sell them for.

I had a strong motivating factor for writing this book in the first place. I wrote it because I think there's a very important need for good, honest and thorough home inspectors. I sincerely want people to be more informed about the realities involved with the biggest investment decision they make - the purchase and sale of their own home. There is also an urgent need to improve the integrity and professionalism of the real estate business overall. By being in the real estate business, I've seen firsthand that there are many aspects about it that need to be improved upon. These improvements would be for the benefit of everyone involved, not just home inspectors. I hope this book will help increase the integrity and professionalism of the real estate business. If it does, then I will feel that this information provides a much needed service and well worth my efforts.

> *I had a strong motivating factor for writing this book in the first place. I wrote it because I think there's a very important need for good, honest and thorough home inspectors. I sincerely want people to be more informed about the realities involved with the biggest investment decision they make - the purchase and sale of their home.*

Please send me an email and let me know what you think of this book and any recommendations you might have for improvements or new products. I accept positive and negative comments since both help me to improve the next version of the book. I am always looking to improve my products and services and I greatly appreciate customer feedback and suggestions. Also, you can sign up through our www.nemmar.com Nemmar Real Estate Training web site to receive our **free** real estate newsletter with articles and product updates that will definitely help you profit in real estate investing.

Purpose Of A Home Inspection

People get very emotional and excited about purchasing a house. When they're in this highly emotional and excited state, they tend to just look at the cosmetic appeal of a house instead of the important factors. They forget that they're not buying a car; they're buying a house! By becoming too emotionally attached to a deal, people often pay above market value for a home. This can cost them tens of thousands of dollars in an overpriced purchase. A house is the biggest investment most people will make, so it's prudent for them not to take any chances. They should try to eliminate as much risk as possible. It's a great feeling to have a client thank you for helping them out with the biggest investment they'll ever make.

> *A house is usually the biggest investment most people will make, so it's prudent for them not to take any chances. They should try to eliminate as much risk as possible.*

People shouldn't buy a house based only on its cosmetic appeal. No house is perfect, there will be repairs or upgrading needed in all homes, even brand new ones. Sometimes people think that because a house is new it doesn't need to be inspected. They don't realize that builders are businessmen trying to make a profit. Any builder who doesn't do quality construction can cut corners to save a few dollars to increase their profit. When a house is built *"up to code"* it doesn't ensure a perfect house. Local building codes are the minimum standards that a builder or contractor has to follow to obtain a building permit or a Certificate of Occupancy for the work done. There's nothing to stop a builder or a contractor from exceeding the building codes other than saving some money for themselves or their client.

A pre-purchase home inspection will inform people of the true condition of a house. This will enable them to make an educated and intelligent decision on whether or not to purchase the home. They will also know what repairs and upgrading will be needed.

Pre-sale home inspections are also recommended. Before someone puts their house up for sale they should have it inspected to find any problems that can be corrected. This will prevent any last minute holdups because of problems found during the buyer's home inspection. Any last minute problems will delay the sale or kill the deal altogether.

As an *"A to Z Home Inspector"* you will be providing a much needed and highly respected service. People are trusting you to help them with the biggest decision they'll ever make!!

Benefits Of Knowledge Of Home Inspections

Being an *"A to Z Home Inspector"* will enable you to be involved in a recession-proof business! Real Estate is still bought and sold during a recession in the economy. In addition to the typical home sales, there are foreclosure sales, relocations, and distressed sales. The only difference about a recession is that houses are sold for a lower price. All of these houses still need to be inspected, and as a result, your business can grow during a slow economy. Unfortunately, there are people who lose their homes due to tragic circumstances in their lives. However, you're not taking advantage of anyone or being unethical by inspecting properties that are the result of a distressed sale or a foreclosure. All of these properties have to be inspected. Therefore, someone is going to be hired to do the job. Everyone will be much better off if the client hires an *"A to Z Home Inspector"* who does top quality work.

Being an *"A to Z Home Inspector"* can make you a much better home buyer or real estate investor. Before you buy a house or a rental property you can check it from top to bottom to figure out what condition it's in. You can save up to tens of thousands of dollars by spotting a problem before it costs you money later. Many home buyers simply pay close to the asking price that the seller lists the house for. They tend to assume that the "listing price" must be close to the market value of the property. Often this is not so. The seller can ask any price they want for their property. However, the true market value could be much less than their asking price. If you have the knowledge of how to perform good, thorough home inspections then you can negotiate some price reduction or concessions. You can intelligently inform the seller if you find any problem conditions during the inspection.

Being an *"A to Z Home Inspector"* can make you a much better home seller. Before you sell your house you can bring everything up to top working condition ahead of time. An awful lot of real estate deals are killed before the real estate appraisal or inspection is done. The buyer and seller may come to an agreement on price and terms pending the home inspection or appraisal. If the home inspector is thorough, they may find problems with the house. This ends up throwing a monkey wrench into the works by lowering the sales price or killing the deal.

Being an *"A to Z Home Inspector"* can make you a much better homeowner. You can make your house safe for you and your family. You'll be able to identify electrical hazards, radon, asbestos, carbon monoxide and gas leaks, etc. Prevent accidents *before* they happen! Also, eliminate building code violations or from being cheated by contractors.

Being an *"A to Z Home Inspector"* can make you a much better Realtor. If you're presently a Real Estate Agent or Broker, then you already know how a low appraisal or problems found during the home inspection can kill a deal. You can greatly increase the percentages of the deals you close. You can show the seller what problem conditions there are at the time you take the listing for the house or condo. As a Realtor, you can inform the home seller yourself without waiting for the buyer's home inspector to do it later.

I'm sure all of you that are presently Realtors are aware of possible complaints. Some people may have bought a house or condo and found out later that there was a problem. They may believe that you knew about the problem or that you should've known about it. As a Realtor, you can help to reduce your headaches and liability by knowing about any potential problems before they come back to haunt you later. You can get more listings due to your knowledge and expertise, which any home buyer or seller will respect. This will enable you to increase your business and income.

Being an *"A to Z Home Inspector"* can make you a much better real estate consultant. If you do any consulting work, you can gain much more respect and business due to your expertise and knowledge.

Being an *"A to Z Home Inspector"* can make you a much better real estate appraiser. I do appraisals as well as inspections and I can see things that 90% of the appraisers in the business don't even know about. My appraisals are much more thorough and informative for the client. Therefore, I can charge a lot more for my work than other appraisers can.

> *So you see, I've got this baby set on automatic pilot. There's something in this material for just about anyone to benefit from.*

So you see, I've got this baby set on automatic pilot. There's something in this material for just about anyone to benefit from. There's only one type of individual that I can think of that can't benefit from this material. That's someone who has no intention of going into the real estate business in any way, shape or form. They also would need to have no intention of ever owning their own house or condo. If that description fits you, then throw this book away right now because it's not for you.

Description Of A Home Inspection

A home inspection is a little different depending on your regional area and the type of housing construction that's more common. However, you'll be dealing with the same general inspection process and business.

A home inspection is a visual, limited time, nondestructive inspection. There's no dismantling or using tools to take things apart. However, you will need some tools to help on the inspection. I'll tell you what tools you need a little further on in this book.

You don't turn anything on that doesn't operate by its normal controls. Meaning that you use only thermostats to test the heating or air-conditioning systems, etc. Don't test anything that has "do-it-yourself" wiring and installations. Any do-it-yourself type of setups can be dangerous to operate and you don't get paid to get injured.

You're a real estate home inspector, you're not a *repairman*. Tell the client to have something checked out by a licensed contractor if it's not operating properly, or if you have any doubts about its present condition. This means that you don't need to know how to fix everything in a house. All you need to do is to be able to identify a problem, or identify whether the operating systems are working properly. To make this point more clear, I'll use an analogy with your car. Normally, you know when your car isn't running properly or if there's a problem with it. However, you don't need to know how to fix the car. All you need to do is identify that something is wrong and that a repairman needs to check it out further. You just bring the car to an auto mechanic and let them tell you exactly what's wrong and what it will cost to repair the problem.

As a home inspector, you're not required to be the **Wizard of Oz**. You're not required to have a *magic wand* that reveals every, single problem with the house and site. You're not required to have *X-ray vision* to see things behind walls, ceilings or other finished areas. You're not required to have a *crystal ball* to foresee **all** potential problems that will arise in the future with the subject property. *(However some people expect you to).*

You determine what the current condition and life expectancy is of different aspects of a house or condominium. Try to be very conservative with evaluating life expectancies of an item because you don't know what the past maintenance history was for the items you're evaluating. As a home inspector, you check all visible and accessible areas and operating systems, such as, heating, air-conditioning, electrical, plumbing, the roof, etc. Everything should be operating properly at the time of the inspection.

A home inspection is not a building code inspection. Building code inspections determine if a property adheres to all of the local building codes. The local municipality has their own building inspectors that perform these services. A home inspection and a building code inspection are two very different services. Home inspectors do not memorize the local building codes for their clients.

Professional Engineer and Registered Architect Issue

For doing home inspections there are licenses and/or certification requirements in most states. Most states generally require you to work under someone else's wing until you learn the basics of the business and get some training. It's similar to getting a State appraiser's license. Working with someone else's inspection company has many positive aspects to it. You won't be calling the shots for a while, but you won't have any overhead or liability problems either.

There are no Engineering or Architecture degrees needed to do home inspections. Let me put this issue to bed nice and early. This is a very common misconception about home inspections. The following statement was issued by the American Society of Home Inspectors. ASHI has members from all different backgrounds, including engineers and architects. This is verbatim from their statement:

> There are no Engineering or Architecture degrees needed to do home inspections.
> Let me put this issue to bed nice and early. This is a very common misconception about home inspections.

"It is not uncommon for buyers to be confused about who is qualified to perform home inspections. There are for example, no home inspection degrees offered or required by law. In some cases, consumers have been led to believe that a home inspection involves engineering analysis and therefore requires the use of a licensed Professional Engineer. The confusion is compounded by the inadvertent misuse of the term "engineer" and "engineering inspection."

Visual home inspections do not involve engineering analysis however even when performed by Professional Engineers. In fact, engineering is an entirely different type of investigation, which entails detailed scientific measurements, tests, calculations and/or analysis. Normally this is done on one specific component of the house (structural or electrical, for example) by, or under the direction of, an engineer trained in that area. Such a technically exhaustive analysis involves considerable time and expense, and is only appropriate on rare occasions when visual evidence exists to indicate design problems which require further, specialized investigation."

An analogy to this is, let's say you were doing an inspection on a slate roof, and there were visible signs that shingles were excessively shaling (flaking) and falling off the house. You should then instruct the client to contact an expert in that field who specializes in that one aspect of house construction. In this example the client should contact a roofer to get repair estimates. A home inspector is a generalist, in the sense that he/she knows a lot about **all** the different aspects of housing construction. A home inspector is not an expert in just **one** area of housing construction.

You have to educate potential clients when they call for price quotes and are only looking for a PE or a RA. PE and RA are the abbreviations for Professional Engineer and Registered Architect. Tell the potential clients to try to find an engineering or architecture school that teaches people how to become a home inspector. *(But don't hold your breath while you look).* Engineering courses are a lot math and physics. Home inspectors are trained to determine the life expectancy and operating conditions of a boiler/furnace, air-conditioning, plumbing, termites, water problems in basements, roofs, septics, wells, etc.

You will come across some unethical engineers and architects that try to deceive people into believing that they're more qualified as home inspectors due to their college degree and education. There was a PE in my area who was suspended from ASHI over this issue. This dishonest PE was caught handing out flyers to Realtors and other people stating that a new law was just passed which stated you must be a Professional Engineer to be a home inspector. This PE even went a step further. He put ads in the yellow pages that misled consumers into believing only Professional Engineers were qualified to do home inspections. Inspectors in the area complained to the yellow pages and they made this PE change his ad the next year. Talk about a greedy, dishonest, jerk! This guy wins the prize.

A PE can be a Nuclear Engineer, a Civil Engineer, a Chemical Engineer, an Aeronautical Engineer, a Mechanical Engineer, an Electrical Engineer, etc. which has absolutely <u>NOTHING</u> to do with a home inspection. Architects are trained to design the plans and layout for homes, not evaluate the life expectancy and condition of roofs, heating systems, plumbing, siding, insulation, water problems, termites, etc. I have friends that are architects, mechanical, electrical, and structural engineers and they know nothing about home inspections. When you come across any engineers or architects who try to tell people that they're the only ones qualified to do home inspections, then you should just ask them what they learned in college about home inspections. Ask them, what Engineering or Architecture school teaches you to become a home inspector? It's not that being a PE or a RA is bad, because it's helpful to have these qualified degrees regardless of the occupation you're in. It's just that a PE or a RA isn't any more qualified to do home inspections than a person with a good construction background or a well-trained home inspector.

You may even want to tell potential clients to call their State Board of Engineering to learn the distinction between an engineer and a home inspector. Do the same with the Board or Architects in your State. If you know what you're talking about and you're knowledgeable enough, the customer will understand the logic behind the PE and RA issue. You should even tell the potential client to question the integrity of any other home inspector that tries to convince them that only PE's or RA's are qualified to do inspections.

Tools That Are Helpful

- Road maps of your area and a car compass to find the job site.
- A clipboard with a notepad and pens to take your field notes.
- Tool box to carry your tools.
- Reliable, powerful flashlight is a necessity.
- Some inspectors like to carry a camera with film to take pictures of the interior, exterior and the operating systems of the house to help them with writing up the inspection report.
- Lighted magnifying glass to view any data plates that are hard to read.
- Large probe and an awl to check wood for rot and termite damage.
- Electric screwdriver to take off any panel covers that are meant for easy removal only.
- Safety glasses to wear in crawl spaces or to view the firebox in a furnace or boiler or other areas.
- Extendable mirror to view furnace or boiler heat exchangers.
- Work gloves if you work in dirty areas or have a dirty furnace or boiler.

- ◊ Hard hat, knee pads and a jump suit to wear in narrow crawl spaces.
- ◊ Voltage tester to test the electrical panel for voltage before touching it.
- ◊ Polarity and GFCI tester to test outlets for proper wiring and operation.
- ◊ Measuring tape that's 16 feet long for any measurements needed for the client.
- ◊ Thermometer to test the air temperature coming from forced hot air and air-conditioning vents.
- ◊ Extendable magnet to reach any screws or small metal parts.
- ◊ Calipers to measure the diameter of a pipe.
- ◊ Combustible gas detector is helpful to test for minor gas and carbon monoxide leaks.
- ◊ A marble and a six inch and a four-foot level to check walls and floors for being level.
- ◊ Pliers to help in some situations, such as lifting the corner of a rug to see the floor underneath.
- ◊ Binoculars to view the roof, chimney, siding and other parts of the house that you can't see clear enough from the ground.
- ◊ Folding ladder to view the roof from close-up.
- ◊ Radon canisters for radon gas testing. It's not so much the canister type, but the quality and sophistication of the radon lab equipment that's important for radon testing. I'll talk more about this later in the book.
- ◊ Septic tank dye to test septic systems.
- ◊ Well water bottles for water laboratory analysis.

Booking Home Inspection Jobs

To give a price quote you have to determine the amount of time and liability that's involved with the inspection. Explain to the client what a prepurchase home inspection is. Let them know it's a visual, limited time, nondestructive and nondismantling inspection. You can't be responsible for things that you can't see, such as, behind finished walls, floors and ceilings. You also can't see any underground systems like wells, septics, oil tanks, etc. Don't scare them off into thinking you won't do anything for them. Just make them realize what a home inspection is. This way everything is up front and they won't think that your inspection is a guarantee that will find all the problems, visible and nonvisible.

I'll list some items that you should find out from the client when giving a price quote for an inspection job. The following items are all listed on the appointment and price quote cards that are included in this book. These index cards will help you give price quotes and keep track of your home inspection appointments. Remember that when giving price quotes, you always have to consider the amount of time the on-site inspection and the written report will take to complete. Another factor to consider is the liability involved.

1. **Is the subject property a condominium, single family, multifamily, etc.?** Condominiums take less time than a single family house inspection. This is because the Condo Association maintains most of the exterior of the condo building. There is a monthly assessment fee charged to all of the individual condo owners to pay for this maintenance. A single family house will take less time to inspect than a multifamily house, especially if the multifamily has a separate heating and/or air-conditioning system for each unit.

2. **What's the square footage and/or the number of bedrooms and bathrooms in the house?** The larger the square footage is, then the longer the inspection will take. Sometimes the client won't know the square footage size of the house so try to find out how many bedrooms and bathrooms there are. For example, a four bedroom and three bath house will be a large home and will take some extra time to inspect.

3. **From what you know, is the house in overall good, average, fair, or poor condition?** The worse condition the house is in, then the longer the inspection will take. Also, if the house is in poor condition and needs a lot of repairs, then there's more risk for you. The client should be able to tell you what the overall appearance of the house is from what they could see. They don't have to be a home inspector to give you some idea of the general condition of the house.

4. **What's the age of the house?** Generally the older the house, the more repairs will be needed and/or there will be some outdated operating systems. This can lead to more risk for you due to the possibility that you miss something that you should have noticed.

5. **Is there a garage?** If the house has a two car detached garage, then you'll spend a little more time evaluating this then if there was no garage.

6. **Is there a basement and/or crawl space?** Basements and crawl spaces can have serious problems in them, especially if the house is older. These areas *must* be inspected thoroughly. You have to account for all of this in your price quote. Also, ask if the basement is finished with wall, ceiling and floor coverings. If it's finished than you won't be able to see behind any inaccessible areas. Make sure your client knows this.

7. **Does the house have a central air-conditioning system?** If it does, then this must be evaluated, *(if the weather is warm enough to test it safely)*. If you include this in the inspection, then you'll spend more time at the site and in writing the report.

8. **Is the house connected to a septic system or the city sewer system?** If the house has a septic system that you're going to dye test, then you should include this in your price quote.

9. **Is the house connected to a well water system or is it supplied by the city water system?** If the house has a well water system that you're going to test, then you should also include this in your price quote.

10. **Do you want a termite and other wood destroying insect inspection?** If the client wants you to check for these insects, then you'll be more liable if you miss something that you should have noticed. So charge the client for this service.

11. **Do you want a radon gas analysis done?** If you test for radon, you want to charge for this service. All houses should be tested due to the health hazards caused by high levels of radon in a home.

12. **Do you want a laboratory water analysis done?** If you test the water, *(you should always test well water)*, for bacteria, mineral and/or radon content then you want to charge for this service also.

13. **Where is the house located?** If the subject property is farther away from your office than the normal inspection site, then you want to charge for the additional traveling time involved. This is important when you start to get busy. While you're away from your office you can't answer the phone to give price quotes. When this happens, you'll miss some jobs, unless of course you hire a secretary to answer your phone.

14. **What is the selling price of the house?** Be careful when asking this since some people don't like telling anyone the sales price when they are buying or selling their home. If people hesitate to tell you the sales price then just explain to them the reason why you are asking that question. It's not that you're trying to be nosy, you just need to know the sales price because the liability risk and time involved to inspect a $1 million dollar home can be far greater than that involved with inspecting a $100,000 dollar home. You have to account for this in your price estimate.

> *If the client asks if you're an engineer or an architect, just educate them about this common misconception.*
> *By telling the facts, you'll earn their respect for being so up-front and honest.*

If the client asks if you're an engineer or an architect, just educate them about this common misconception. By telling them the facts, you'll earn their respect for being so up-front and honest.

Tell the client that it's highly recommended that they attend the inspection. This will enable them to see firsthand all of the different aspects of the house you'll be evaluating. Having the client attend the inspection also helps to eliminate questions, phone calls and problems later. Tell them that it's also recommended that they arrange the inspection at a time when the owner of the house you'll be inspecting will be home. The reason for this is that there are many questions you need to ask the owner of the house directly. As an **"A to Z Home Inspector"** you need to ask these questions to obtain some information to help you with the home inspection. I'll explain more about this later.

Sometimes you'll book jobs to inspect vacant houses. Some houses are left vacant when being sold for a number of different reasons. The homeowner could have died and it's an estate sale; the owner may have been relocated by his company for a new job position; the owner may be away for a long vacation;

it could be a bank foreclosure sale, etc. If the subject property is vacant, then there are important items to be aware of. Often, vacant houses will have the utilities turned off. You should notify the client of this when booking the job. I've arrived at houses many times to do a home inspection or appraisal and the utilities were turned off. This limits what you can evaluate. For example, without electricity you can't check the outlets and switches; without gas or oil you can't test the boiler/furnace or water heater; without the water supply turned on you can't test the plumbing pressure and drainage. There's another aspect to be aware of with vacant houses. If the property is located in cold weather areas, then the heating system must be kept on all winter or else the water pipes must be winterized. This protects the pipes from water freezing, expanding and cracking the pipes.

Pre-inspection contracts are starting to gain support among many inspection companies. The purpose of these is to have the client sign a contract before the inspection. The contract is designed so that the client will understand what the inspection involves and what the limitations of it are.

There are inspection companies that offer some of the home warranty programs that are on the market. Home warranty programs offer the home buyer a type of insurance policy. The buyer obtains the insurance so that if they buy the house and something breaks down or there's a problem, they may be reimbursed for any expenses. This is different than Errors and Omissions insurance. E and O insurance covers the home **inspector**. Home warranty insurance covers the home **buyer**. If you're going to offer a warranty policy to your clients, then make sure that you read the fine print and that you understand them *completely*. Sometimes these policies are very limited in their coverage protection, so you and your client need to know up front what your client will be getting for their money.

Many of these home warranty policies only offer the client a depreciated value reimbursement for any claims. They also do not cover certain aspects of the house and the coverage period is limited to about 12 months after the client moves in. This means that if the client buys the house and the boiler needs replacing in 14 months, then the policy will not cover this expense. Also, when a claim is paid it's usually depreciated. This is similar to auto insurance. When an insurance adjuster "totals" a used car after an accident, the insurance company only pays you the book value of that car. They don't buy you a new car! Basically it's up to you. Some people feel it's a selling point to offer their clients a warranty policy. It's pretty much a judgment call from your own perspective. So look into the warranty policies in your area and decide if there's one for you. But whatever you decide, *make sure you read the fine print* so your client doesn't think he's getting "full blanket" insurance coverage.

There are some seller disclosure forms that have come out for home sellers to sign when they're marketing their house. You will come across some people who try to convince home buyers that they don't need to get the house inspected. They tell the home buyers that an inspection isn't needed because of a seller disclosure form and/or a warranty program.

The seller disclosure forms are **very limited** because the seller of a home knows *nothing* about home inspections. The seller can only tell the home buyer if things are working up to their standards, which may be different from the buyer's standards. For example, the seller may have no problem living in a house with low water pressure or an occasional water problem in the basement. However, your client who is the home buyer, has totally different needs and standards that they're looking for in their purchase. This can't be evaluated properly by a seller disclosure. Also, let's say the seller disclosure states that the roof has no water leaks. This could be a true statement. Maybe the roof isn't leaking now. But what if that roof is 20 years old and it's going to need replacing within a year? The seller knows **nothing** about roofs. So how can the seller tell a buyer that there's no need to worry about the roof. These are the reasons why a home inspection is still needed, even if there are a seller disclosure form and a warranty program for the house.

It's very important that your client understands the severe limitations of the home warranty programs and the seller disclosure forms.
Your client has a right to know this information.

It's very important that your client understands the severe limitations of the home warranty programs and the seller disclosure forms. Your client has a *right* to know this information. So don't let any Realtors, sellers or other third parties try to convince them that they don't need to get the house inspected.

Beginning The Home Inspection

Some areas of the country, like Florida, don't have basements. As you move from the lower level through the interior and up to the attic, move in a clockwise direction. This will help prevent you from bouncing around from room to room which may cause you to skip a room by accident. I'll always start the home inspection in the lower level because this is usually where the operating systems are located. I usually spend at least one hour in the lower level of a house inspecting the operating systems and for structural, water and termite problems. I start the inspection with the heating system in the winter or the air-conditioning in the summer. If it's late in the afternoon, I'll start with the exterior before it gets dark. Then I can take my time on the interior.

You can modify the home inspection process to meet your own needs or desires. You'll be nervous for the first ten or so home inspections. This is normal. Just remember that you need to learn this material well enough, and keep up to date with all the new construction trends. If you do then you will earn the respect of the buyer and all third parties to the transaction by being so knowledgeable.

When I say third party people, I'm talking about people involved in the transaction, not including your client. This could be any number of people. The list includes but isn't limited to: the seller, the Realtor, the bank appraiser, the mortgage lender, the attorney, the seller's dog or cat, and anyone else who has an interest in the deal.

If you're young, sometimes people will get a little worried when they first meet you at the job site. I have no idea, but for some strange reason some people seem worried about a home inspector who's young. That is, until they see that you know what you're talking about.

When driving up to the house or condo you should take note of the condition of it, the terrain, if there are any ponds or streams, etc. Mark down the time the inspection begins and ends. Mark down the weather conditions. Any snow covered areas will not be visible for inspection. If it is/was raining there may be signs, or lack of signs, of water in the lower level and any roof leaks.

Greet the owner and Realtor and just tell them you have to ask some questions about the house or condo to get some background. You need this info to help you with the report and the home inspection. There are some aspects of the house that you can't always detect or verify without some additional information from the seller or Realtor. Often you'll find that you can't get all the information you need from the questions you ask the owner or Realtor. Just get whatever information you can and keep a record of it. Make sure that you put their answers in the written report to CYA, which stands for Cover Your Assets. *(You can remove some letters off the end of Assets if you like)*. This will help in the event that you find out later that someone misrepresented the house or condo. You'll be able to show proof about what was stated and represented to you and your client at the time of the inspection. This is why you want to stress to the client to arrange the inspection at a time when the owner of the house is home. It's important to tell your client this when you're booking the inspection. This way they'll have time to notify the owner to arrange the appointment. You should also get a copy of any real estate listing sheets, surveys, etc. See if there's anything important in these documents to help you or your client.

You have to be very gentle when you ask the seller of the house the items listed in the **Questions To Ask The Home Seller** section on page 29. Sometimes they get very upset and worried about all these questions. Just tell them that it's nothing personal or that you don't trust them, you just need this information to assist you with the inspection. There are many aspects about a house that only the owner may know about and that's what you're trying to find out. If they (the seller) were buying the house, they would want you to find out the same information from the seller as well. **Just remember that you're a guest in someone else's house! So don't be rude or get into an argument with anyone at the inspection.** You must be diplomatic and professional in this or any other business to be successful.

Definition Of Terms

The following definitions are general guidelines to use when writing up your home inspection report or filling out your on-site checklist at the subject property. When you describe the condition of a particular item at the subject property you want to make sure you use the same criteria in your definitions for all items so your inspection report is consistent. Using the following guidelines will simplify the process for you.

GOOD = The condition and/or appearance of the subject item is *above* the condition normally expected for a similar item. There are no significant repairs and upgradings needed at the present time in the visible and accessible areas. The age, apparent quality of manufacture, exposure to the elements, reported maintenance history, etc., are all taken into account in evaluating the subject item.

AVERAGE = The condition and/or appearance of the subject item is *equal to* the condition normally expected for a similar item. There are no significant repairs and upgradings needed at the present time in the visible and accessible areas. Normal maintenance is expected for the subject item. The age, apparent quality of manufacture, exposure to the elements, reported maintenance history, etc., are all taken into account in evaluating the subject item.

FAIR = The condition and/or appearance of the subject item is *below* the condition normally expected for a similar item. There are some significant repairs and upgradings needed at the present time in the visible and accessible areas. Above normal maintenance is expected for the subject item. The age, apparent quality of manufacture, exposure to the elements, reported maintenance history, etc., are all taken into account in evaluating the subject item.

POOR = The condition and/or appearance of the subject item is *far below* the condition normally expected for a similar item. There are significant repairs and upgradings needed at the present time in the visible and accessible areas. Far above normal maintenance is expected for the subject item. The age, apparent quality of manufacture, exposure to the elements, reported maintenance history, etc., are all taken into account in evaluating the subject item.

NOTE: There will be abbreviations next to some of the applicable blank lines on the checklist pages to assist you in filling out the home inspection checklist at the subject property. The abbreviations are:

◊ (G/A/F/P) = Good/Average/Fair/Poor

◊ (y/n) = Yes/No/Not. You can put a check mark ✓ for a "yes" answer. Some items can have more descriptive answers like "slight", "some", "average", "a lot" "abnormal", "extensive", "excessive", etc.

◊ (#) = Enter the applicable number or amount for this item.

Sample Inspection Report Pages

The Operating Systems Inspection 27

The Operating Systems Inspection

Heating System

◊ The brand name of the heating system installed is __Bradford Furnace__.

◊ The fuel for the heating system is (gas/oil/electricity) __oil__.

◊ The overall BTU capacity of the heating system is __180,000__.

◊ The age of the heating system is approximately (#) __14__ years. The life expectancy is generally (#) __23__ years for this type of heating system.

◊ The last date of service/repairs for the heating system was __December__.

◊ The ceiling and walls around the heating system should have a covering of sheet-metal or 5/8 inch fireproof sheetrock to help prevent the spread of fires in this area. The fireproof covering is (y/n) __not__ installed.

◊ The flue pipe sections are in (G/A/F/P) __Average__ condition. The joints at the connecting sections do (y/n) __✓__ have the required screws to keep them in place. The flue pipe does (y/n) __✓__ have the required upward pitch. These are all required items that must be installed. The flue pipe is used to safely discharge the carbon monoxide and other products of combustion that are caused by gas and oil fired burners. These gases **must be safely discharged** from the house. They are *LETHAL GASES!!*

◊ The flue pipe is (y/n) __not__ within 4 inches of any combustible material, such as wood. A minimum clearance of 4 inches is required to help prevent fires.

◊ The heating system is operated by (#) __2__ zone(s). A zone is an area of the house with a separate thermostat which can have a different setting. Having more than one heating zone is more energy efficient.

◊ The heating system was tested by turning up all zone thermostats to engage the heating system for about (#) __35__ minutes. All radiators/registers did (y/n) __✓__ operate properly during the testing by getting warm.

◊ The emergency shutoff switch did (y/n) __not__ operate properly when tested. This is used to shut the system off during repairs and for emergencies by overriding the thermostat control.

- There is some rust noted at the base of the Furnace.
- Determine from the seller if there is a service contract for the Furnace and get copies of all oil bills.
- Make repairs as noted above for safety.

The Operating Systems Inspection

◊ **Gas Fired Heating Systems:**

◊ There is (y/n) __y__ a required gas shutoff valve within 6 feet of the burner for safety to shut off the fuel. The purpose of this is to be able to shut off the flow of gas if it is necessary. This is different from the emergency shutoff switch because it does not turn off the burner, it only shuts off the gas supply.

◊ The gas feed lines do (y/n) __y__ appear to be made of approved piping. The gas lines MUST have approved black iron gas piping for the feed lines and not copper or other materials that are not approved to carry gas fuel.

◊ The burner flames were (y/n) __not__ able to be checked for proper color and height. They should be checked periodically and should be as blue as possible with very little yellow or orange color. Too much yellow or orange color means that the fuel and air mixture needs to be adjusted.

◊ The inspector was (y/n) __not__ able to view the heat exchanger. There were (y/n) __some__ signs of excessive rust or cracks in the visible and accessible areas. Sealed systems cannot be fully inspected due to lack of access.

◊ The draft diverter hood at the base of the flue pipe is in (G/A/F/P) __Fair__ condition. It is used to keep downdrafts in the chimney from blowing out the pilot light and to help keep the heat inside the boiler or furnace while the carbon monoxide is removed.

- The burner and heat exchanger areas were sealed. Determine if these areas have been serviced by the heating contractor.
- Monitor the rust noted on the exterior of the heat exchanger area.
- Update the draft diverter hood for safety.
- Gas shutoff valve was stuck and could not be turned off which is a safety hazard. Repairs are needed.

◊ **Electric Heating Systems:**

◊ The Owner or Realtor stated that there is (y/n) _____ problem with blown fuses or tripped circuit breakers due to the additional electrical usage for the electrical heating system.

◊ All of the electric baseboard radiators were (y/n) _____ warm during the testing. The radiators are in (G/A/F/P) _____ condition.

(Not Applicable)

Domestic Water Heater

◊ **Separate Domestic Water Heaters:**

◊ The domestic water heater is fueled by (gas/oil/electricity) __Gas__.

◊ The capacity of the water heater is (#) __80__ gallons. The standard size water heater for a single family house is 40 gallons. Oversized water heaters in the house are not as energy efficient because a lot of water will just sit in the tank and will not be used after being heated.

◊ There were (y/n) __✓__ signs of excessive rust or water leaking conditions on the unit. Any rust or leaking conditions will require repairs by a licensed contractor.

◊ The temperature/pressure relief valve is (y/n) __not__ located directly on the water heater for safety. This valve is in (G/A/F/P) __Good__ condition. This is a safety device that helps prevent the water heater from becoming dangerously high in temperature or pressure. If the temperature reaches 210 degrees Fahrenheit or the pressure reaches 150 psi, then the valve will discharge to relieve the system so the water heater will not explode.

◊ The pressure relief valve is (y/n) __not__ piped properly for safety. It *must* be piped to within 8 inches of the floor to prevent scalding anyone when discharging. When water or steam discharges from this valve, it indicates a problem condition and the system *must* be checked out immediately by a licensed contractor.

◊ There is (y/n) __✓__ a drain valve on the lower part of the water heater. This is used to drain some water into a bucket each month to remove some of the rust and sediment that normally builds up in the system.

◊ The hot and cold water lines do (y/n) __not__ appear to have a reversed installation *(hot pipe in cold pipe slot)*. A reversed installation with the hot and cold lines will cut down the energy efficiency of the water heater.

◊ The estimated age of the water heater is (#) __15__ years. The life expectancy of a water heater is 10 to 12 years depending upon the maintenance given to it over the years.

◊ The water heater temperature setting was (#) or (hot/warm/cold) __warm__ at the time of the inspection. The water heater should be kept on the *"warm"* setting for maximum efficiency and life expectancy. A high temperature setting will cause the unit to constantly heat the water higher than is necessary, which can cause excessive wear and tear and premature failure.

◊ During the interior inspection the faucets were spot checked and adequate hot water was (y/n) __✓__ available.

- The water heater is old, very rusty, and oversized! Replace this with a new 40 gallon water heater.
- The temp/pressure relief valve is new and was replaced recently - BUT it is not located directly on the water heater tank - it's on the water pipe, and it's not piped properly! These are safety hazards!
- The "warm" setting is noted to be 125° according to the heater data label.

Attic Inspection

◊ The attic space was (y/n) *partly* accessible for the inspector to make evaluations of the areas above the top floor finished ceilings.

◊ The access panel leading to the attic is located *in ceiling of upper hallway*. There is (y/n) *pulldown* stairway to provide easy access to the attic. The stairway was in (G/A/F/P) *Average* condition and did (y/n) *✓* have sturdy handrails and steps for safe passage.

◊ It is <u>highly</u> recommended that a handrail be installed inside the attic area surrounding the access opening. This will help prevent anyone walking in the attic area from falling through the access opening. Do not wait for an accident to happen, install a handrail <u>now!!</u>

◊ The attic floor is (y/n) *partly* covered so that this area can be used to store **lightweight** household items. Heavy objects are not recommended for storage in the attic area due to the excessive weight they exert on the ceiling below.

◊ The roof ridge beam appeared to be in (G/A/F/P) *Good* condition where visible and accessible. The roof *"ridge beam"* is the main girder type beam at the top of the crest of the roof.

◊ The roof rafters appeared to be in (G/A/F/P) *Average* condition where visible and accessible. The roof *"rafters"* are the floor joist type beams leading from the attic floor up to the roof ridge beam.

◊ The roof sheathing appeared to be in (G/A/F/P) *Poor* condition where visible and accessible. The roof *"sheathing"* is the sub-flooring type wood that the roof shingles rest upon. The sheathing is made of (plywood/etc) *plywood - but there are signs of old water damage.*

◊ There were (y/n) *excessive* water stains that appeared to be due to water leaks or abnormal humidity in the attic area. Often there are old water stains from prior roof leaks that have been repaired. There are (y/n) *no* bowed and/or damaged sections of the wood members in the attic roof area. Any water leaks, abnormal humidity, bowing and/or damaged wood members in the attic roof area will require repairs by a licensed roofing contractor.

◊ There are (y/n) *✓* collar beams in the attic. *"Collar beams"* are generally constructed of 2 x 4 inch wood boards that are located several feet below the ridge beam. The purpose of collar beams is to *"tie"* both sides of the roof together so that all of the weight of the roof does not rest upon the ridge beam alone. Collar beams give the roof additional support and it is recommended that they be installed when not noted.

- Attic pull-down stairs were replaced last year according to the Owner of the house.
- There are excessive water stains because the roof leaked before the prior Owner replaced the roof shingles according to the Seller. The plywood roof sheathing should be replaced when the next roof is needed on the house. This is an expensive repair when done so budget for it.

The Exterior Home Inspection

Roof

◊ There are (y/n) _a few_ tree branches overhanging the roof area. There should not be any tree branches overhanging the roof, which can cause damage to the shingles.

◊ The inspector was (y/n) _✓_ able to view the roof adequately from the following location(s) _front and rear of the house and along the left side_. All judgments and evaluations are made from that perspective.

◊ There are (y/n) _no_ bowing sections of the roof ridge beam, roof rafters and/or the roof sheathing which would indicate repairs being needed.

◊ The roof covering/shingles installed on the house is made of _asphalt/Fiberglass and was installed 2 years ago_. The roof covering/shingles appears to be approximately (#) _2_ years old. It is in (G/A/F/P) _GOOD_ condition. There are (y/n) _no_ signs of rapidly aging and damaged areas of the roof covering.

◊ The average life expectancy of this type of roof covering/shingles is (#) _18-22_ years. The life expectancy of all roofs depends upon many factors, some of which are the quality of the roof covering, the quality of the installation, the climate and exposure to the elements and the maintenance given to the roof over the years.

◊ The slope of the roof that has a southerly or southwesterly exposure faces the sun more often and can become brittle and show signs of aging faster. The slope of the roof that has a northerly or northeastern exposure is more apt to have mold and decay fungi due to the lack of sunlight.

◊ There appears to be (#) _2_ layer(s) of shingles on the roof. Some local building codes allow there to be up to 3 layers of shingles on a roof. However, this adds too much weight to the structure and it is only recommended that there be 2 layers as a maximum. Also, when shingles are placed over an existing layer they tend to have a poor cosmetic appearance and it cuts down their life expectancy. When there are 2 or more layers of roof covering installed presently, then you will have to strip these layers off during the next re-roofing. This is much more expensive than just having a new layer put over the existing shingles. *Especially* if the roof sheathing has to be replaced as well.

- Determine if there is a warranty with the roof shingles that applies to the home buyer.
- When the roof needs replacing, the sheathing will need removing as noted in the Attic area.
- Some mold was noted on the front, left shingles due to the overhanging tree branches. Prune them away.

Subject Property #1

◊ The address of the subject property is _____.

◊ The inspection was commissioned by _____.

◊ The date of the inspection was _____.

◊ The time of the inspection was from _____ to _____ AM PM.

◊ The outdoor air temperature was approximately _____ degrees.

◊ The weather on the day of the inspection was (raining/sunny)_____.

◊ *<u>You must use and read the entire inspection report to get the maximum benefit!</u>* Do not just take it for granted that everything is in good condition at the subject property. The written inspection report has valuable and important information that you need to know in order to properly evaluate the subject property.

◊ A home inspection is a visual, limited time, non-destructive, and non-dismantling inspection. There is no dismantling or using tools to take things apart. There are areas that are inaccessible or not visible, such as behind finished wall, floor and ceiling coverings, etc.

◊ A home inspection checks to see if all visible and accessible areas and operating systems, such as, heating, air-conditioning, electrical, plumbing, roof, etc. are operating properly at the time of the inspection. The inspection tries to determine what the current condition and life expectancy is of the different aspects of the subject property. The inspection is limited in evaluating the life expectancies of an item without knowing what the past maintenance history was for the item being evaluated.

◊ During the inspection you cannot turn on or test any devices, appliances, operating systems, electrical switches, etc., that do not operate by the normal controls used to operate it that were designed by the manufacturer. For example, you can only use thermostats to test the heating or air-conditioning systems, etc. You cannot test anything that has *"do-it-yourself"* wiring and installations. Any do-it-yourself type of setups can be dangerous to operate and are not part of the home inspection process.

◊ It is *highly* recommended that any problem conditions noted at the on-site inspection or in this written report be evaluated by a reputable, licensed contractor **PRIOR** to signing any contracts or closing on the subject property. The inspection performed is not a building code inspection. It is also recommended that you check <u>all</u> records at town hall pertaining to the subject property.

Questions To Ask The Home Seller #1

When you ask these preinspection questions, make sure that you ask the owner and Realtor about information from any prior owners of the house. Meaning that if the seller tells you,
"No, we have never made any changes to the foundation or septic system,"
then ask them if they know of any <u>prior</u> owners having made any changes, repairs, etc.

◊ *Age, Zoning, and Permits:*

◊ What is the age of the house/condo? _____

◊ How long you lived in this house/condo? _____

◊ Are there any outstanding building, zoning or other violations or any missing permits and/or approvals?
_____

◊ *Interior Inspection:*

◊ Are there any damaged areas to the floors, walls, and/or ceilings that you know about? _____

◊ If yes, then where are these damaged areas and are they hidden by carpets, furniture, sheetrock, etc.?
_____

◊ Has any insulation been added or removed in the floors, walls, and/or ceilings? _____

◊ If yes, then explain the details of the insulation added or removed: _____
_____

◊ Has any UFFI foam or asbestos insulation been removed from the house and if yes, then do you have the licensed EPA contractor certification for that work? _____

◊ Does the fireplace draft properly or are there back-smoking problems and how often do you use the fireplace?
_____

◊ *Exterior Inspection:*

◊ Has there been any exterior siding added to the house after the original construction? _____
_____

◊   Do you know what type of insulation and materials are behind the exterior siding between the walls?
_____

◊   What is the age of the roof? _____

◊   Do you know how many layers of shingles there are on the roof? _____

◊   Have there been any past or present water leaks or problems with the roof? _____
_____

◊   Have any decks, additions or updating been done? If yes, what are the details of that work and are all valid permits and Certificate of Occupancies, *(C of O)*, filed at town hall? _____
_____

◊   Have any structural renovations been done? If yes, is there a valid permit and C of O for the work done?
_____

◊   Have there been any structural problems in the house? If yes, what are the details? _____
_____

◊   **<u>Operating Systems:</u>**

◊   Can I test all operating systems in the house or are there any that are being repaired or aren't functioning properly? *(Operating Systems refers to items such as the heating, air-conditioning, plumbing, electrical, wells, septics, etc.)* _____
_____

◊   Do you know the age of the furnace/boiler? _____

◊   How often and what company services/maintains the heating system? _____

◊   Are all the rooms in the house heated? _____

◊   Are there any oil tanks, used or unused, on the property? _____

◊   If yes, where are those oil tanks located and what is the age of those oil tanks? _____
_____

◊   Have there been any problems with the heating system in the house? If yes, what are the details? _____
_____

◊   Do you know the age of the air-conditioning compressor? _____

◊   Did the air-conditioning system operate properly last season? *(if it's too cold to test it now)* _____

◊ How often and what company services/maintains the A/C system? _____

◊ Have there been any problems with the air-conditioning system in the house? If yes, what are the details? _____

◊ Is the house/condo connected to municipal water & sewer systems? *(This is very important to get from them since there is no way to determine this at the site without checking the town hall records).* _____
_____

◊ Has there been any past or present problems with the water pressure and drainage in the house plumbing system? _____
_____

◊ Has there been any past or present problems with electrical overloads, outlets, switches, etc.? _____
_____

◊ ***Termite and Water Problems:***

◊ Has the house ever had termites or Wood Destroying Insect damage? _____

◊ Has the house ever been treated to remove or prevent termites or Wood Destroying Insects? _____

◊ If yes, when and what are the details of that WDI treatment? Are there any guarantees or documentation for the WDI treatments? _____
_____

◊ Have there been any water penetration problems in the house? If yes, what are the details? _____
_____

◊ Are there any sump pumps in the lower level area to remove water? _____

◊ ***Septic System:***

◊ Is there a survey or plot plan showing the septic system? _____

◊ Have there been any renovations or additions to the house needing septic system approvals, such as bathrooms/bedrooms added? _____

◊ Have there been any past or present problems with the septic system? _____
_____

◊ Do you know the location of the septic tank and leaching fields? _____

◊ Do you know what the size of the septic tank is? _____

◊ Do you know what construction materials the septic tank is made of? _____

◊ Is the septic tank original or was it upgraded or replaced over the years? _____

◊ When was the septic tank last pumped out and internally inspected? _____

◊ How often has the septic tank been pumped out and internally inspected over the years prior to the last cleaning? _____

◊ What is the name of the septic service company that maintains the system? _____

_____

◊ *Well Water System:*

◊ Are there any surveys or plot plans showing the well water system? _____

◊ What is the depth of the well? _____

◊ Is the well water pressure and volume adequate for normal use? _____

◊ Has there been any past or present problems with the well water pressure and volume? _____

_____

◊ When was the well pump last serviced or replaced? _____

◊ When was the well water storage tank last serviced? _____

◊ What is the age of the well water storage tank? _____

◊ What is the name of the well water service company that maintains the system? _____

_____

◊ *Swimming Pool:*

◊ What is the age of the pool, filter, heater and liner? _____

◊ Do you have a Certificate of Occupancy and all valid permits for the pool? _____

◊ Have there been any leaks in the pool walls or other problems with the pool or pool equipment? _____

◊ Has the pool been properly winterized? *(if applicable)* _____

◊ What is the name of the service company that maintains the swimming pool? _____

_____

_____

The Operating Systems Inspection #1

Heating System

◊   The brand name of the heating system installed is _____.

◊   The fuel for the heating system is (gas/oil/electricity)_____.

◊   The overall BTU capacity of the heating system is _____.

◊   The age of the heating system is approximately (#)_____ years. The life expectancy is generally (#)_____ years for this type of heating system.

◊   The last date of service/repairs for the heating system was _____.

◊   The ceiling and walls around the heating system should have a covering of sheet-metal or 5/8 inch fireproof sheetrock to help prevent the spread of fires in this area. The fireproof covering is (y/n)_____ installed.

◊   The flue pipe sections are in (G/A/F/P)_____ condition. The joints at the connecting sections do (y/n)_____ have the required screws to keep them in place. The flue pipe does (y/n)_____ have the required upward pitch. These are all required items that must be installed. The flue pipe is used to safely discharge the carbon monoxide and other products of combustion that are caused by gas and oil fired burners. These gases **must be safely discharged** from the house. They are ***LETHAL GASES!!***

◊   The flue pipe is (y/n)_____ within 4 inches of any combustible material, such as wood. A minimum clearance of 4 inches is required to help prevent fires.

◊   The heating system is operated by (#)_____ zone(s). A zone is an area of the house with a separate thermostat which can have a different setting. Having more than one heating zone is more energy efficient.

◊   The heating system was tested by turning up all zone thermostats to engage the heating system for about (#)_____ minutes. All radiators/registers did (y/n)_____ operate properly during the testing by getting warm.

◊   The emergency shutoff switch did (y/n)_____ operate properly when tested. This is used to shut the system off during repairs and for emergencies by overriding the thermostat control.

_____
_____
_____
_____

◊ *<u>Oil Fired Heating Systems:</u>*

◊   Oil fired systems *<u>must</u>* be tuned up every year by a reputable heating service contractor for efficient operation. Most oil delivery companies will provide a service contract with the Owner to service and tune up the oil burner and provide emergency repairs and maintain the oil tanks. The burner flame must be adjusted every year, the oil filters changed, the flue draft regulator adjusted, and the flue pipe cleaned.

◊   When the oil burner engaged, there were (y/n)_____ signs of back-smoking. Any back-smoking will indicate a problem condition that needs to be repaired.

◊   The oil burner flame was (y/n)_____ able to be viewed for proper color and height.

◊   The firebox is in (G/A/F/P)_____ condition. This area needs to be periodically monitored due to deterioration from the high temperature of the burner flame.

◊   The oil feed lines are in (G/A/F/P)_____ condition. The feed lines should be made of copper and should be covered to prevent them from being damaged or becoming a tripping hazard.

◊   There is (y/n)_____ a required firematic shutoff valve within 6 feet of the burner for safety to shut off the fuel. The purpose of this is to be able to shut off the flow of oil if it is necessary. This is different from the emergency shutoff switch because it does not turn off the burner, it only shuts off the oil supply.

◊   The oil filter is in (G/A/F/P)_____ condition.

◊   The oil tank is located _____. The oil tank was (y/n)_____ able to be viewed for rusting or corrosion problems. It is *<u>highly</u>* recommended that the oil tank be tested by a reputable oil contractor to determine if there are any present leaks or potential leaks due to corroded sections of the tank.

◊   Oil tanks generally last about 25 to 30 years and longer if maintained properly. Determine from the Owner or the oil contractor what the exact age of the oil tank(s) is. Determine if any Certificate of Occupancy, permits or surveys are needed in the local municipality with on-site oil tanks.

◊   The draft regulator on the flue pipe is in (G/A/F/P)_____ condition. This needs to be adjusted every year with the tune-up of the heating system. The purpose of it is to allow some cool air from the boiler room area to help with the removal of carbon monoxide up the chimney, without drawing too much heat from the boiler or furnace.

_____
_____
_____
_____
_____
_____

The Operating Systems Inspection #1　　　　　　　　　　　　　　　　　　　　　　　www.nemmar.com   35

◊ **_Gas Fired Heating Systems:_**

◊ There is (y/n)_____ a required gas shutoff valve within 6 feet of the burner for safety to shut off the fuel. The purpose of this is to be able to shut off the flow of gas if it is necessary. This is different from the emergency shutoff switch because it does not turn off the burner, it only shuts off the gas supply.

◊ The gas feed lines do (y/n)_____ appear to be made of approved piping. The gas lines **MUST** have approved black iron gas piping for the feed lines and not copper or other materials that are not approved to carry gas fuel.

◊ The burner flames were (y/n)_____ able to be checked for proper color and height. They should be checked periodically and should be as blue as possible with very little yellow or orange color. Too much yellow or orange color means that the fuel and air mixture needs to be adjusted.

◊ The inspector was (y/n)_____ able to view the heat exchanger. There were (y/n)_____ signs of excessive rust or cracks in the visible and accessible areas. Sealed systems cannot be fully inspected due to lack of access.

◊ The draft diverter hood at the base of the flue pipe is in (G/A/F/P)_____ condition. It is used to keep downdrafts in the chimney from blowing out the pilot light and to help keep the heat inside the boiler or furnace while the carbon monoxide is removed.

_____

_____

_____

_____

_____

_____

◊ **_Electric Heating Systems:_**

◊ The Owner or Realtor stated that there is (y/n)_____ problem with blown fuses or tripped circuit breakers due to the additional electrical usage for the electrical heating system.

◊ All of the electric baseboard radiators were (y/n)_____ warm during the testing. The radiators are in (G/A/F/P)_____ condition.

_____

_____

_____

◊ **_Forced Warm Air Systems:_**

◊ There is (y/n)_____ at least one supply vent in each room providing heat.

◊ The air filters were (y/n)_____ clean when checked. They need to be replaced every few months during the heating season. This is similar to changing your car air and oil filters. If you do not do it often enough, you will create excess wear and tear on the furnace due to the lack of maintenance.

◊ The air ducts are (y/n)_____ insulated. All ducts should be insulated, internally or externally, for maximum energy efficiency.

◊ There is (y/n)_____ a humidifier on the heating ducts. Forced hot air heating systems will dry out the air in a house and can lead to the occupants getting sore throats. A humidifier will help prevent this. Humidifiers cannot be fully tested during a home inspection due to their operation. The life expectancy of a humidifier is generally about 5 to 7 years.

◊ The furnace plenum is (y/n)_____ separated from the heat exchanger and fan area by a canvas or flexible type of material. This will help prevent any vibrations caused by the blower fan from being transmitted through the ducts and into the livable rooms.

◊ The inspector was (y/n)_____ able to view the heat exchanger and the fan. There were (y/n)_____ signs of excessive rust or cracks in the visible and accessible areas. Sealed systems cannot be fully inspected due to lack of access. If the heat exchanger is cracked it will leak carbon monoxide into the supply ducts of the house. If there is a leak of any kind it must be checked out **_IMMEDIATELY_** before using the heating system again. The carbon monoxide that is released is a **LETHAL GAS!!**

◊ The fan was (y/n)_____ making unusual noises while it was operating. Abnormal or unusual noises while operating will indicate repairs may be needed.

_____

_____

_____

_____

_____

_____

_____

_____

_____

◊ *Heat Pump Heating Systems:*

◊ The Owner or Realtor stated that the heat pump unit does (y/n)_____ heat the house adequately in the cold months.

◊ Heat pumps cannot be tested in both the heating and the air-conditioning modes during an inspection. Doing so can damage the compressor unit. When it is working properly in one mode, then it is an indication that the most important and costly parts are operating properly.

◊ The age of the compressor unit is approximately (#)_____ years. The life expectancy of a compressor is about 10 to 12 years, depending upon the amount of usage and maintenance given to the system over the years.

◊ The compressor is (y/n)_____ resting on a sturdy, level foundation, like concrete. Uneven installations can cause premature failure of the compressor because it will be leaning to one side while operating.

◊ The compressor was (y/n)_____ making unusual noises while it was operating. Abnormal or unusual noises while operating will indicate repairs may be needed.

◊ There is (y/n)_____ a required exterior service disconnect switch next to the compressor for emergency and repairs shut off.

◊ There was (y/n)_____ adequate ventilation around the compressor unit for the air intake and blower fan to operate properly. All trees and bushes should be pruned away at all times and there should be no obstructions overhead.

◊ The compressor coils were (y/n)_____ clean when checked. The coils need to be cleaned periodically for proper maintenance and operation of the system.

_____
_____
_____
_____
_____
_____
_____
_____
_____
_____
_____

◊ **_Forced Hot Water Heating Systems:_**

◊ There is (y/n)_____ at least one radiator in each room for heating purposes.

◊ The circulator pump(s) is in (G/A/F/P)_____ condition and was (y/n)_____ operating properly at the time of the inspection. Circulator pumps need just a drop of oil in the oil ports approximately once a year. Have the heating service contractor check this with the annual tune ups.

◊ The visible heating pipes and pipe joints are in (G/A/F/P)_____ condition. There are (y/n)_____ signs of excessive rust or leaking conditions. Any rust or leaking conditions will require repairs by a licensed contractor.

◊ The heat exchanger was (y/n)_____ able to be viewed for signs of excessive rust or leaking conditions.

◊ The water pressure reducing valve is in (G/A/F/P)_____ condition. This reduces the water pressure that is coming from the house plumbing lines, which is usually about 30 to 60 psi *(pounds per square inch)*, down to about 12 to 15 psi before entering the boiler.

◊ There is (y/n)_____ a required backflow preventer next to the water pressure reducing valve. This prevents water that has entered the boiler from re-circulating back into the house plumbing lines and mixing with the faucet and shower water supply which is a health hazard.

◊ The expansion tank is in (G/A/F/P)_____ condition. When you heat water it expands. The heated water needs a cushion to expand or else it will burst some of the pipe joints or discharge the pressure relief valve to relieve the pressure. The expansion tank has an air pocket or a rubber bag in it that cushions the water as it expands so the pressure in the system does not get too high. It needs to be checked periodically for water-logging problems. It should have a bleeder valve to put air in it if it becomes waterlogged. It should have a drainage valve on the bottom to drain it at least once a year for any rust and sediment that builds up in the tank.

◊ The pressure gauge does (y/n)_____ appear to be operating properly. The proper operating pressure for a hot water heating system is between 12 to 22 psi.

◊ The pressure relief valve is (y/n)_____ located directly on the boiler for safety. This valve is in (G/A/F/P)_____ condition. This is a safety device that helps prevent the heating system from becoming dangerously high in pressure. If the pressure reaches 30 psi then the valve will discharge to relieve the system pressure so the boiler will not explode.

◊ The pressure relief valve is (y/n)_____ piped properly for safety. It _must_ be piped to within 8 inches of the floor to prevent scalding anyone when discharging. When water or steam discharges from this valve, it indicates a problem condition and the system _must_ be checked out immediately by a licensed heating contractor.

◊ There is (y/n)_____ a drain valve on the lower part of the boiler. This is used to drain some water into a bucket each month to remove some of the rust and sediment that normally builds up in the system.

_____

_____

_____

_____

◊ **_Steam Heating Systems:_**

◊ There is (y/n)_____ at least one radiator in each room for heating purposes.

◊ The visible heating pipes and pipe joints are in (G/A/F/P)_____ condition. There are (y/n)_____ signs of excessive rust or leaking conditions. Any rust or leaking conditions will require repairs by a licensed contractor.

◊ The heat exchanger was (y/n)_____ able to be viewed for signs of excessive rust or leaking conditions.

◊ The pressure gauge does (y/n)_____ appear to be operating properly. The proper operating pressure for a steam heating system is between 2 to 5 pounds per square inch, *(psi)*.

◊ The upper limit switch is (y/n)_____ located directly on the boiler for safety. This switch is in (G/A/F/P)_____ condition. If the pressure gets too high, this switch will turn off the burner.

◊ The pressure relief valve is (y/n)_____ located directly on the boiler for safety. This valve is in (G/A/F/P)_____ condition. This is a safety device that helps prevent the heating system from becoming dangerously high in pressure. If the pressure reaches 15 psi then the valve will discharge to relieve the system pressure so the boiler will not explode.

◊ The pressure relief valve is (y/n)_____ piped properly for safety. It must be piped to within 8 inches of the floor to prevent scalding when discharging. When water or steam discharges from this valve, it indicates a problem condition and the system *must* be checked out immediately by a licensed heating contractor.

◊ There is (y/n)_____ a drain valve on the lower part of the boiler. This is used to drain some water into a bucket each month to remove some of the rust and sediment that normally builds up in the system.

◊ The water level in the boiler sight glass was (y/n)_____ at the proper level at the time of the inspection. The sight glass water level should be 1/2 to 3/4 of the way full. It should not be completely empty nor completely full which will cause problems in the system. The sight glass level allows you to see that there is air and water in the boiler. This is because you are making steam and you have got to have room for the heated water to boil and create the steam.

_____
_____
_____
_____
_____
_____
_____
_____

Air-Conditioning System

◊   There is (y/n)_____ a central air-conditioning system for the subject property.

◊   The Owner or Realtor stated that the unit does (y/n)_____ cool the house adequately in the warmer months. An air-conditioning system is a closed system, and, theoretically, there should never be a need for additional Freon. However, in practice, the various fittings on the connecting pipes can loosen or develop hairline cracks that can allow some of the Freon gas to escape. If the air-conditioning system cannot hold a Freon charge for at least one season, then the leaks in the pipes or fittings should be located and sealed.

◊   The outdoor air temperature was (y/n)_____ warm enough to test the air-conditioner at the time of the inspection. Central or window air-conditioning units **cannot** be tested when the outdoor air temperature is 65 degree Fahrenheit or lower. The interior pressure that is required to properly operate an air-conditioning system is too low when the outdoor air temperature is 65 degrees or lower.

◊   All window and wall air-conditioning units were (y/n)_____ able to be spot checked for proper operation. They did (y/n)_____ operate properly when tested. Determine from the Owner or Realtor if the portable air-conditioning units are being sold with the house.

◊   The air-conditioning system was tested by turning up all of the zone thermostats to engage the system for about (#)_____ minutes. All vents/registers did (y/n)_____ operate properly during the testing by discharging cool air.

◊   The inspector was (y/n)_____ able to spot check a supply vent with a thermometer to determine if the discharging air was cool enough. The reading should be about 55 to 58 degrees Fahrenheit. The temperature reading noted was (#)_____ degrees at the time of the inspection.

◊   There is (y/n)_____ at least one supply vent in each room providing cool air.

◊   The inspector was (y/n)_____ able to view the air filters. The air filters were (y/n)_____ clean when checked. They need to be replaced every few months during the air-conditioning season. This is similar to changing your car air and oil filters. If you do not do it often enough, you will create excess wear and tear due to the lack of maintenance.

◊   The air ducts are (y/n)_____ insulated. All ducts should be insulated, internally or externally, for maximum energy efficiency.

◊   The air handler plenum is (y/n)_____ separated from the fan area by a canvas or flexible type of material. This will help prevent any vibrations caused by the blower fan from being transmitted through the ducts and into the livable rooms.

◊   The inspector was (y/n)_____ able to view the evaporator coil and the fan. There were (y/n)_____ signs of excessive rust or cracks in the visible and accessible areas that need to be evaluated by a licensed contractor. Sealed systems cannot be fully inspected due to lack of access.

◊   The inspector was (y/n)_____ able to view the condensation drain pan under the evaporator coil. The condensation drain pan is in (G/A/F/P)_____ condition. The drain pan drainage line does (y/n)_____ lead to a condensate pump that removes the condensation to a suitable location. The life expectancy of these pumps is about 5 to 7 years.

◊   The fan was (y/n)_____ making unusual noises while it was operating. Abnormal or unusual noises while operating will indicate repairs may be needed.

◊ The age of the compressor unit is approximately (#)_____ years. The life expectancy of an air-conditioning compressor is about 10 to 12 years, depending upon the amount of usage and maintenance given to the system over the years.

◊ The compressor is (y/n)_____ resting on a sturdy and level foundation, such as a concrete base. Uneven installations can cause premature failure of the compressor because it will be leaning to one side while operating.

◊ The compressor was (y/n)_____ making unusual noises while it was operating. Abnormal or unusual noises while operating will indicate repairs may be needed.

◊ There is (y/n)_____ a required exterior service disconnect switch next to the compressor for emergency and repairs shut off. There was (y/n)_____ adequate ventilation around the compressor unit for the air intake and blower fan to operate properly. All trees and bushes should be pruned away at all times and there should be no obstructions overhead. The coils were (y/n)_____ clean when checked. The coils need to be cleaned periodically for proper maintenance of the system.

◊ The high and low pressure lines were checked where visible and accessible. The lines are (y/n)_____ made of copper. The low pressure line, which is the larger pipe that is about 3/4 inch in diameter, is (y/n)_____ insulated. This line must be insulated for energy efficiency since it has cold Freon in it. The high pressure line is the thin diameter line and it does not need to be insulated.

Domestic Water Heater

◊ *__Separate Domestic Water Heaters:__*

◊ The domestic water heater is fueled by (gas/oil/electricity)_____.

◊ The capacity of the water heater is (#)_____ gallons. The standard size water heater for a single family house is 40 gallons. Oversized water heaters in the house are not as energy efficient because a lot of water will just sit in the tank and will not be used after being heated.

◊ There were (y/n)_____ signs of excessive rust or water leaking conditions on the unit. Any rust or leaking conditions will require repairs by a licensed contractor.

◊ The temperature/pressure relief valve is (y/n)_____ located directly on the water heater for safety. This valve is in (G/A/F/P)_____ condition. This is a safety device that helps prevent the water heater from becoming dangerously high in temperature or pressure. If the temperature reaches 210 degrees Fahrenheit or the pressure reaches 150 psi, then the valve will discharge to relieve the system so the water heater will not explode.

◊ The pressure relief valve is (y/n)_____ piped properly for safety. It _must_ be piped to within 8 inches of the floor to prevent scalding anyone when discharging. When water or steam discharges from this valve, it indicates a problem condition and the system _must_ be checked out immediately by a licensed contractor.

◊ There is (y/n)_____ a drain valve on the lower part of the water heater. This is used to drain some water into a bucket each month to remove some of the rust and sediment that normally builds up in the system.

◊ The hot and cold water lines do (y/n)_____ appear to have a reversed installation *(hot pipe in cold pipe slot)*. A reversed installation with the hot and cold lines will cut down the energy efficiency of the water heater.

◊ The estimated age of the water heater is (#)_____ years. The life expectancy of a water heater is 10 to 12 years depending upon the maintenance given to it over the years.

◊ The water heater temperature setting was (#) or (hot/warm/cold)_____ at the time of the inspection. All water heaters should be kept on the "warm" setting for maximum efficiency and life expectancy. The warm setting on the water heater thermostat is usually about 125-130 degrees Fahrenheit which is the factory recommended temperature setting for most water heaters. A high temperature setting will cause the unit to constantly heat the water higher than is necessary, which can cause excessive wear and tear and premature failure.

◊ During the interior inspection the faucets were spot checked and adequate hot water was (y/n)_____ available.

_____
_____
_____
_____
_____

◊ ***Immersion Coil Water Heaters:***

◊   The domestic hot water is supplied by an immersion coil system inside the boiler. An immersion coil system has water pipes that carry cold water inside of a coil located in the side of the boiler. The coils are *"immersed"* in the hot boiler water, hence you get the name *"immersion coils."* The cold water in the pipes **should not** mix with the boiler water because then the dirty boiler water would be carried back to the faucets and showers, which would be a health problem.

◊   The advantage of an immersion coil system is that it provides inexpensive hot water in the winter time because the boiler is already operating to heat the house. The disadvantage of an immersion coil system is that it is not as energy efficient as having a separate water heater unit in the warmer months. It also adds an unwanted heat load on the house in the warmer months. Also, the coils clog over time and need to be cleaned periodically.

◊   During the interior inspection the faucets were spot checked and adequate hot water was (y/n)_____ available.

◊ ***Oil Fired Water Heaters:***

◊ Oil fired units have a very fast recovery rate, which is the rate at which they can re-heat the water. There is (y/n)_____ a required water temperature setting switch. The purpose of this is to operate like a thermostat to regulate the burner to turn on and off to keep the water at a pre-set temperature. The factory recommended setting is usually at 125-130 degrees Fahrenheit. If it is set too high, then someone can get scalded with very hot water.

_____
_____
_____
_____
_____

◊ ***Gas Fired Water Heaters:***

◊ There is (y/n)_____ a required water temperature setting switch. The purpose of this is to operate like a thermostat to regulate the burner to turn on and off to keep the water at a pre-set temperature.

_____
_____
_____
_____
_____

◊ ***Electrically Operated Water Heaters:***

◊ Electrically operated units have coils that are directly immersed in the water inside the tank. They usually have two switches inside the small cover plates on the side of the tank to regulate the water temperature setting.

_____
_____
_____
_____
_____

Plumbing System

◊ The visible plumbing lines and joints were in (G/A/F/P)_____ condition. There were (y/n)_____ signs of excessive rust or leaks that need to be further evaluated by a licensed plumbing contractor. The types of plumbing line materials used in housing construction include: Copper, Brass, Galvanized Iron, Lead, PVC, and Cast Iron. The visible **supply** plumbing lines are made of _____. The visible **drainage** plumbing lines are made of _____.

◊ The visible portion of the water main entry line is made of _____. The main water line is (y/n)_____ securely fastened to the wall to help prevent damage to the plumbing joints.

◊ The water main shutoff valve is in (G/A/F/P)_____ condition. This valve does (y/n)_____ appear to be in proper working order. This valve is used to turn off **all** of the water entering the house in case of an emergency or if any repairs are being performed. The main shutoff valve cannot be tested during a home inspection due to the possibility of it becoming rusty over time and if tested it can *"freeze"* in the on or off position, or possibly leak.

◊ There is (y/n)_____ a water meter reading device installed on the main water line. This is used by the utility company to calculate the water usage bill for the homeowner.

◊ There is (y/n)_____ an electrical grounding wire on the water main line for the electrical system. **This is a very important safety item and must not be disconnected or rusty!!** The grounding wire and clamps **must** be checked periodically for any rust or corrosion. The purpose of this wire is to ground the electrical system for safety. The grounding wire does (y/n)_____ span the water meter and shutoff valve. The grounding wire should be clamped on both sides of the water meter and the shutoff valve with a *"jumper cable"* for safety.

◊ There is (y/n)_____ a pressure reducing valve installed by the water meter. This valve is generally found in areas where the municipal water system has very good pressure from the street water lines.

◊ During the interior inspection, the inspector spot checked the water pressure and drainage by briefly running the faucets. The water pressure and drainage was (y/n)_____ in proper working order. The client was (y/n)_____ present to view this testing.

_____

_____

_____

_____

_____

_____

_____

Well Water System

◊   There is (y/n)_____ a well water system for the subject property.

◊   The inspector is very limited in evaluating a well system because most of the components are underground and/or not visible, such as the well pump, water lines and sometimes even the water storage tank is located in an underground pit. Also, the inspector has no accurate knowledge of what the past maintenance history has been for the well and the well equipment.

◊   It is *highly* recommended that you obtain all building department permits, surveys, plot plans and approvals for the well system. Most wells are deep and the repair costs and fees to drill deep wells are **much higher** than shallow wells because the price is usually based on the length of water piping used and the drilling depth.

◊   It is *highly* recommended that you have a water sample taken and analyzed at a reputable laboratory whenever the house has a well water system or possible lead plumbing lines. Laboratory water analysis should be done to test for bacteria, mineral, metal and radon content in the water supply.

◊   The well pump is located inside the well and is not visible. The age of the well pump is reported to be (#)_____ years. The life expectancy of a well pump is about 7 to 10 years, but can be longer if it is not overworked or neglected. The life expectancy also depends upon the type and quality of the pump installed and the acidity of the well water.

◊   The well pressure gauge does (y/n)_____ appear to be operating properly. Pressure gauges often get rusty and need to be replaced every few years.

◊   The water storage tank is in (G/A/F/P)_____ condition. There were (y/n)_____ signs of excessive rust on the tank that need to be evaluated further by a licensed well contractor. The tank is (y/n)_____ insulated to help prevent rust from condensation. The age of the storage tank is approximately (#)_____ years. The life expectancy of a water storage tank is about 15 to 20 years, and similar to a well pump, it depends upon the maintenance done, the type and quality of the tank and the acidity of the well water.

◊   There is (y/n)_____ a pressure relief valve for the well storage tank. This *must* be installed for safety in case the pressure in the system gets too high. It is usually set at 75 psi, *(pounds per square inch),* depending upon the type and capacity of the storage tank. The tank does (y/n)_____ have an air fill valve to adjust the air-to-water ratio inside of it during periodic maintenance of the system.

◊   The visible well water lines are in (G/A/F/P)_____ condition. There were (y/n)_____ signs of excessive rust on the lines. The lines are (y/n)_____ insulated to help prevent rust from condensation.

◊   The well water system was tested by turning on (#)_____ faucets to engage the system for about (#)_____ minutes. All faucets did (y/n)_____ operate properly during the testing by providing an adequate flow of water. The water flow at the time of the inspection was approximately (#)_____ gallons per minute. The minimum acceptable flow for a well system is 5 gallons per minute, *(GPM)*. Some local area codes may require a higher GPM rating. The client was (y/n)_____ present to view the testing.

◊   The *"idle"* or *"static"* pressure reading was (#)_____ psi before the well testing began. During the well water test the pressure gauge had a high reading of (#)_____ psi and a low reading of (#)_____ psi. The well pressure should remain within a 20 psi differential during and after the test. This simply means that the high and low pressure gauge reading of the system should not be more than a 20 psi difference during use.

◊   There is (y/n)_____ a properly operating emergency shutoff switch for the well pump.

◊ There is (y/n)_____ a water filter system installed. There is (y/n)_____ a water softener installed. The inspector cannot evaluate water filtration or water softener systems during a home inspection because of the laboratory water analysis that would be needed. The water filters and brine need to be replaced according to the owners manuals and whenever they appear dirty.

◊ If a water softener is installed and the house is serviced by a septic system, then the brine water from the softener should not discharge into the septic system. The brine water will alter the natural bacterial action of the septic decomposition inside the septic tank and can cause premature failure of the septic system.

Septic System

◊ There is (y/n)_____ a septic system for the subject property.

◊ The inspector is very limited in evaluating a septic system because most of the components are underground and/or not visible, such as the drainage lines, the holding tank and the leaching fields or seepage pits. Also, the inspector has no accurate knowledge of what the past maintenance history has been for the septic system.

◊ It is *highly* recommended that you obtain all building department permits, surveys, plot plans and approvals for the septic system. The life expectancy of a septic system is about 30 years depending upon the type of construction materials used and the maintenance given to it over the years. The repair costs and costs to re-build or move a septic system are **very high**!!

◊ It is *highly* recommended that you have a licensed septic contractor pump clean and internally inspect the septic system prior to closing. The dye testing performed during a home inspection is **very limited** and does not always reveal a septic system that has failed or is on the verge of failure. When the system is pumped out clean, the septic contractor can *internally* inspect the holding tank and the drainage lines coming into and out of it. This gives him a visual look at the interior of the tank and often the septic contractor will provide a written report for this service. Another benefit for you to get the septic system pumped clean prior to closing, is that if you do buy the house, then they will be moving in with a cleaned out septic tank that should not need any maintenance for quite some time. A septic contractor can also partially dig up the leaching field area to do a more extensive evaluation. This will allow them to determine if the leaching fields or septic drainage pipes are clogged.

◊ Septic systems **MUST** be cleaned at least every 2 to 3 years to properly maintain them. It should be more frequent than every two years if there are a lot of people in the house or if the homeowner does a lot of entertaining and often has guests/parties at the home. It is highly recommended that you obtain all documentation of the past septic maintenance records for future use. Many times a neglected septic system will be pumped clean just because the house is being put on the market for sale. It is similar to driving a car for many years without changing the oil. The car will run on dirty oil, but it will cost you money in wear and tear and it will eventually die prematurely due to the lack of maintenance given to it.

◊ The Owner or Realtor stated that the date the septic system was last pumped clean was _____. They also stated that the septic system was pumped out and cleaned approximately (#)_____ years before this last cleaning.

◊ The septic system was tested by turning on (#)_____ faucets for about (#)_____ minutes. All toilets were flushed (#)_____ times during the testing. A harmless, colored dye designed for septic testing was flushed down the toilets at the beginning of the septic test. The dye did (y/n)_____ show up on or around the property. If the dye is seen or there is a septic odor in the lawn during the testing, then this indicates a problem condition with the septic system and a licensed contractor needs to evaluate the septic system.

_____

_____

_____

_____

_____

Electrical System

◊ The electrical system service entrance line is located _____.
There are (y/n)_____ tree branches touching the electrical lines and equipment. All tree branches near the lines and equipment can cause damage and should be pruned away for safety.

◊ The electrical service entrance head is in (G/A/F/P)_____ condition. There were (y/n)_____ signs of excessive rust or corrosion on the areas that are visible to the inspector.

◊ There was (y/n)_____ a drip loop on the electrical lines before they enter the service entrance head. A *"drip loop"* is created by slack in the wiring in a "U" shape. This helps prevent rainwater from following the electrical lines down into the main panel.

◊ There are (#)_____ service entrance lines going into the house. Two service entrance lines indicate that there is 110 volts inside. A 110 volt electrical system will usually only have a maximum of 30-60 amps of electrical service in the main panel. Three service entrance lines indicate that there is 220 volts inside. A 220 volt electrical system can have up to 200 amp electrical service in the main panel.

◊ The exposed electrical service lines from the service entrance head to the main electrical panel were (y/n)_____ enclosed in a conduit. A *"conduit"* is a covering to protect the electrical lines from the weather and damage. The conduit is in (G/A/F/P)_____ condition and there were (y/n)_____ signs of cracked or open areas and/or joints that are not sealed properly.

◊ The electrical meter is located _____. When the meter is located on the exterior, then the utility company can take a reading without having to enter the house.

◊ **Remember that electricity can kill you!!** Before touching the main panel or any sub-panels check them with a voltage tester to make sure that they are not electrified. Do not go near any exposed wiring or any electrical panels or wiring if there is water on the floor. Water and electricity *don't* mix!!

◊ The main electrical panel is located _____. There were (y/n)_____ signs of excessive rust or corrosion on the main electrical panel. Any sign of excessive rust or corrosion requires a licensed electrician to properly evaluate the electrical system for safety. The main electrical panel is (y/n)_____ installed on the wall securely. There are (y/n)_____ hazardous conditions around the panel, such as, potential water, objects in the way, the panel being too high to reach safely, etc.

◊ There are (y/n)_____ sub-panels noted. *"Sub-panels"* are small electrical panels that branch off from the main electrical panel. The purpose of sub-panels is to prevent very long branch circuit runs in the house.

◊ The electrical system has (circuit breakers/fuses)_____ for the branch circuits. The inspector cannot turn any circuit breakers off or on or replace any fuses, for safety reasons. There are (y/n)_____ tripped circuit breakers or blown fuses in the main electrical panel and/or sub-panels. Any *"tripped"* circuit breakers or *"blown"* fuses indicate a problem that <u>must</u> be evaluated by a licensed electrician.

◊ All circuits are (y/n)_____ marked to indicate where each branch circuit leads to. This is a convenience and safety feature. The markings will assist the homeowner in turning off individual branch circuits in case of an emergency or if repairs are needed. There is **no way** for the inspector to determine if the circuits are properly marked for the exact location in the house for their corresponding branch lines.

◊ There are (y/n)_____ open circuit breaker or fuse slots in the main panel and/or sub-panels. Open slots need to be covered with *"blanks"* or spare circuit breakers or fuses. This will prevent anyone from sticking their fingers or any objects inside the electrical panel and getting electrocuted.

◊   There is (y/n)_____ room in the main panel for additional branch circuits. Any unused circuits will generally allow the homeowner to expand the system by adding more branch circuits directly from the main panel without having to install sub-panels.

◊   The main electrical disconnect is located _____. This will shut off **all** of the electrical current leading into the house from the main panel. The main disconnect is (y/n)_____ installed at a safe height that is readily accessible. The main disconnect should be at least 30 inches above the ground and no more than 7 feet high for safety. This will enable it to be safely turned off in case of an emergency.

◊   The amperage for the electrical system is determined to be (#)_____ amps. The National Electrical Code, *(NEC)*, recommends that the minimum amperage be 100 amps for a residential property. The **main disconnect switch** does (y/n)_____ have a visible amperage rating number. The **main electrical panel** does (y/n)_____ have a visible amperage rating number. The **service entrance lines** are (y/n)_____ visible to be viewed for the voltage capacity leading into the house. All three of these indicators must be visible to the inspector to accurately determine the amperage of the electrical system.

◊   The electrical grounding wire is (y/n)_____ installed properly and attached to (water main/grounding rod)_____. There are (y/n)_____ signs of excessive rust or corrosion that require repairs by a licensed electrician. **It is extremely important that the electrical system be grounded to a properly working grounding cable that is attached to the water main line or a grounding rod that extends at least 10 feet into the soil.** The grounding wire and clamps **must** be checked periodically and must not be disconnected or rusty. The purpose of this is to ground the electrical system for safety.

◊   There are (y/n)_____ signs of loose and/or exposed electrical wiring. There are (y/n)_____ signs of loose electrical switches and outlets that need to be secured. Any loose and/or exposed wiring, switches and outlets *must* be secured by a licensed electrician to prevent any electrical hazards.

◊   There are (y/n)_____ signs of *"do-it-yourself"* or unprofessional electrical work. All electrical repairs must be performed by a licensed electrician and all valid permits and building department approvals must be obtained for any work performed.

◊   The outlets and switches are (y/n)_____ reachable from the bathroom tubs or showers. As a safety precaution the outlets and switches in the bathroom **must not** be reachable from the tub or shower. Remember that water and electricity do not mix!!

◊   A limited electrical outlet tester was used to spot check the outlets for proper wiring and current. There are (y/n)_____ outlets noted with improper wiring and/or grounding. Improper wiring can be caused by having the hot and neutral wires reversed in the back of the outlet or a false ground wire. Outlets with these conditions may still provide electrical current but they are an electrical safety hazard and *must* be repaired by a licensed electrician.

◊   There are (y/n)_____ two pronged outlets noted that *must* be upgraded to modern three pronged outlets by a licensed electrician. Older houses will have two pronged outlets as opposed to the modern three pronged type. The third prong is used for the grounding prong in electrical cord plugs. The purpose of this grounding prong is that most appliances today have an internal ground for electrical safety reasons.

◊   There are (y/n)_____ properly operating Ground Fault Circuit Interrupters, *(GFCI)*, noted in some of the outlets and/or in the electrical panel. A *"GFCI"* is an electronic device that will *"trip"* (turn off) the circuit when it senses a potentially hazardous condition. It is very sensitive and operates very quickly. The quick response time in interrupting the power is fast enough to prevent injury to anyone in normal health. GFCI's are recommended by the National Electric Code to be installed anywhere near water for safety. Such water prone areas are basements, garages, kitchens, bathrooms and all exterior outlets.

The Operating Systems Inspection #1   www.nemmar.com

◊ There are (y/n)_____ an adequate number of electrical outlets noted. There were (y/n)_____ electrical extension cords in use at the time of the inspection. The NEC recommends that houses have an outlet for every 6 feet of horizontal wall space. This is because most appliances come with 6 foot electrical cords and if there are not enough outlets, then the homeowner will have to use extension cords. Extension cord wiring is not recommended because of the possibility of someone plugging an appliance into a low amperage rated extension cord. This will cause the extension cord wire to overheat and start an electrical fire.

◊ It is *highly* recommended that the homeowner install child proof electrical outlet caps if there are any children in the house. These are small plastic plugs to cover any unused outlets so a child will not stick anything into the outlets and get electrocuted.

_____
_____
_____
_____
_____
_____
_____
_____
_____
_____
_____
_____
_____
_____
_____
_____

Additional Comments

The Lower Level Inspection #1

Lower Level

◊ There is (y/n)_____ a lower level area for the subject property.

◊ The lower level is (y/n)_____ finished with areas that are not accessible or not visible. The lower level has (y/n)_____ areas that are inaccessible due to personal items and furniture of the homeowner, or lack of access for the inspector to view areas. Any inaccessible areas cannot be evaluated by the inspector.

◊ The lower level stairs are in (G/A/F/P)_____ condition. The stairs do (y/n)_____ have sturdy handrails with closely spaced posts to prevent tripping hazards and children from falling through the railing openings. The stairs do (y/n)_____ have evenly spaced steps to help prevent tripping hazards.

◊ The house construction materials used for the foundation walls, in the visible areas, are made of _____.

◊ The floor of the lower level does (y/n)_____ have a concrete covering. A concrete floor is recommended in the lower level areas to help prevent moisture and wood destroying insect problems in the house. The floor is in (G/A/F/P)_____ condition where visible and accessible.

◊ There are (y/n)_____ signs of abnormally large settlements cracks in the visible areas of the walls and floors of the lower level. Any **long, horizontal settlement cracks** or any cracks that are over 1/4 of an inch wide **MUST** be further evaluated by a licensed contractor to determine the possible cause and repair options.

◊ There are (y/n)_____ signs of areas of the foundation that have been altered from the time of the original construction of the house. Any alterations to the structure from the time of the original construction will require valid permits and approvals from the municipality for the work performed.

◊ The main girder beam(s) was checked, where accessible, and was in (G/A/F/P)_____ condition and is made of _____. The *"main girder"* of a house is the large beam that spans across the entire width of the house. This is the beam that supports the interior portions of the house and it rests on the top the foundation walls. There are (y/n)_____ signs of excessive rusting, rotted, cut-out, cracked and/or sagging sections of the main girder beam(s). When any of these conditions are noted, the beam(s) **must** be evaluated by a licensed contractor for safety.

◊ The support posts were checked, where accessible, and are in (G/A/F/P)_____ condition and are made of _____. The *"support posts"* of a house are found underneath the main girder beam(s) and should be spaced about 6 feet apart. These posts support the main girder beam(s) in the middle sections of the house lower level area while the ends of the main girder beam(s) are supported by the foundation walls. There are (y/n)_____ signs of excessive rusting, rotted, cut-out, cracked and/or sagging sections of the support posts. When any of these conditions are noted, the posts **must** be evaluated by a licensed contractor for safety.

◊ The floor joists were checked, where accessible, and are in (G/A/F/P)_____ condition. The *"floor joists"* are the wood boards that span across the underside of the floors in the house, which in turn hold up the floors. The floor joists run perpendicular to the main girder beam(s). The floor joists do (y/n)_____ have the required diagonal bracing installed in the visible areas. These are small wood boards or metal straps placed diagonally in between the floor joists to *"tie"* them together for additional strength. There are (y/n)_____ signs of excessive rotted, cut-out, cracked and/or sagging sections of the floor joists. When any of these conditions are noted, the joists **must** be evaluated by a licensed contractor for safety.

◊ The sub-flooring was checked, where accessible, and was in (G/A/F/P)_____ condition. The *"sub-flooring"* is the plywood boards or paneled wood boards that are located on top of the floor joists. The purpose of the sub-flooring is to support the finished flooring above, such as hardwood, tiles or carpeting, that rests on top of the sub-flooring. There are (y/n)_____ signs of excessive rotted, cut-out, cracked and/or sagging sections of the sub-flooring.

_____
_____
_____
_____
_____
_____
_____
_____
_____
_____
_____
_____
_____
_____
_____
_____
_____
_____
_____
_____

Crawl Spaces

◊ There is (y/n)_____ a lower level crawl space area for the subject property.

◊ The inspector did (y/n)_____ have adequate access to view the crawl space area. *"Crawl spaces"* are small areas underneath the livable portions of the house which are not high enough to stand up in. This is an area that **demands** attention periodically since there is a higher risk of rot and termite infestation due to it being dark and damp most of the time.

◊ There does (y/n)_____ appear to be adequate ventilation in the crawl space area. Crawl spaces need plenty of ventilation all year round to help prevent rot and wood destroying insect infestation.

◊ The crawl space floor has a (dirt/concrete/etc)_____ covering. Any crawl spaces that have a dirt floor should be covered with a concrete surface. Dirt floors will promote moisture from the soil and are an attraction to wood destroying insects. If putting concrete over the dirt floor is too expensive then a 6 mil plastic floor cover can be placed over the dirt areas to help eliminate some of the moisture problems in the crawl space.

◊ There are (y/n)_____ signs of abnormally large settlements cracks in the visible areas of the walls and floors of the crawl space. Any **long, horizontal settlement cracks** or any cracks that are over 1/4 of an inch wide **MUST** be further evaluated by a licensed contractor to determine the possible cause and repair options.

◊ There are (y/n)_____ signs of areas of the foundation in the crawl space area that have been altered from the time of the original construction of the house. Any alterations to the structure from the time of the original construction will require valid permits and approvals from the local municipality for the work performed.

◊ The main girder beam(s) of the crawl space was checked, where accessible, and was in (G/A/F/P)_____ condition and is made of _____. There are (y/n)_____ signs of excessive rusting, rotted, cut-out, cracked and/or sagging sections of the main girder beam(s). When any of these conditions are noted, the beam(s) **must** be evaluated by a licensed contractor for safety.

◊ The support posts of the crawl space were checked, where accessible, and are in (G/A/F/P)_____ condition and are made of _____. There are (y/n)_____ signs of excessive rusting, rotted, cut-out, cracked and/or sagging sections of the support posts. When any of these conditions are noted, the posts **must** be evaluated by a licensed contractor for safety.

◊ The floor joists of the crawl space were checked, where accessible, and are in (G/A/F/P)_____ condition. The floor joists do (y/n)_____ have the required diagonal bracing installed in the visible areas. There are (y/n)_____ signs of excessive rotted, cut-out, cracked and/or sagging sections of the floor joists. When any of these conditions are noted, the joists **must** be evaluated by a licensed contractor for safety.

◊ The sub-flooring of the crawl space was checked, where accessible, and was in (G/A/F/P)_____ condition. There are (y/n)_____ signs of rotted, cut-out, cracked and/or sagging sections of the sub-flooring.

_____

_____

_____

_____

Gas Service

◊   The subject property reportedly is (y/n)_____ connected to local gas utility lines in the street. When there is no gas service in the building, you should determine from the local utility company whether or not it is available and what costs are involved in having gas service installed in the house. Installing gas service can be a major expense to the homeowner.

◊   The gas meter is located _____. The gas meter and visible gas lines are in (G/A/F/P)_____ condition. There are (y/n)_____ signs of excessive, loose and/or leaking gas lines. The visible gas lines are made of _____ and do (y/n)_____ appear to be an approved type of gas piping. All gas service lines should be approved black iron gas piping. Any loose, leaking, excessive rusting conditions, or unapproved types of metals being used for gas feed lines **must** be evaluated by a licensed contractor to make any necessary repairs to bring the gas lines up to the building codes.

◊   There is (y/n)_____ a main shut-off valve near the gas meter for safety. This is used to shut-off the gas supply for repairs or in case of an emergency. If you smell or detect any gas leaks in the house, **immediately** contact the local utility company to make repairs. **Leaking gas will explode so *do not* take any chances!!**

◊   There is (y/n)_____ Liquid Petroleum Gas *(LPG)* tank(s) noted on the site at the time of the inspection. The LPG gas tank(s) is in (G/A/F/P)_____ condition. There are (y/n)_____ signs of excessive rust or corrosion. The tank(s) does (y/n)_____ appear to be properly level on a sturdy foundation. Any problem condition **must** be checked out a reputable LPG contractor for safety. It is recommended that you check with town hall to make sure all valid permits and approvals are on file for the existence of any LPG tank(s) on the site.

◊   **Do not** bring any gas tanks into the house, such as, exterior barbecue tanks or automobile gas cans. Barbecue gas tanks are under extreme pressure, similar to scuba diving tanks. *If they ever exploded they could blow up the entire building and everyone in it!!*

Auxiliary Systems

◊ There is (y/n)_____ an auxiliary system(s) for the subject property.

◊ Auxiliary systems, such as, burglar alarm systems, fire detection systems, intercoms, central vacuum systems, lawn sprinklers, etc., *are not* evaluated during a limited time home inspection. Obtain all manuals from the Owner and find out how to operate these systems, when they are present on the site. Determine from the Owner or Realtor if any fire or alarm systems are hooked up to any monitoring services and/or the local police or fire departments and what the fees are for this service.

◊ The following auxiliary systems were present on the site:

_____

_____

_____

Water Penetration

◊ Signs of water penetration can be white mineral salts on the concrete walls and floors. This is called *"efflorescence"* and is caused by water seeping through the concrete and then drying on the exterior portion leaving the white, mineral salt from the cement as a residue.

◊ Most lower level areas will get some minor efflorescence on the lower portion of the walls and floors due to normal humidity in this area because it is underground. It is recommended that you use a dehumidifier to help prevent moisture in any lower level areas.

◊ Sometimes in the corners there may be indications of water stains. Often this is caused by the lack of gutters and downspouts on the house or downspouts that drain right next to the foundation walls on the exterior. All downspouts should be piped at least 5 feet away from the house so the rainwater will not drain next to the foundation and then enter the lower level of the house. Sometimes the downspouts drain into underground drainage lines. These lines can become clogged due to leaves or a small animal getting stuck in them. Gutters, downspouts and underground drainage lines need to be checked periodically for proper operation.

◊ The grading of the soil next to the exterior of the house can also cause water stains on the lower level walls and floors. All soil next to the foundation should be slightly sloped away from the side of the house to help prevent rainwater from entering the lower level.

◊ There are (y/n)_____ signs of abnormal or excessive water problems in the house beyond normal humidity and condensation stains.

◊ The accessible wood members that are in contact with the floor, such as, workbench posts, storage items, etc., were probed to reveal any rotting conditions or evidence of abnormal or excessive water penetration. The wood members that were spot checked are in (G/A/F/P)_____ condition and did (y/n)_____ have signs of abnormal or excessive water penetration.

◊   There is (y/n)_____ a sump pump located in the lower level. These are small pumps that help carry water away from the house. Sump pumps are located in small pits dug into the lower level floor and have a drainage pipe to carry water to a more desirable location.

◊   The sump pit walls are in (G/A/F/P)_____ condition. Any *"do-it-yourself"* or other unprofessional installations need to be repaired by a licensed contractor. There was (y/n)_____ water inside the sump pit at the time of the inspection. Any water in the sump pit indicates that the area has a high groundwater table and that there is a potential for water to enter the lower level of the house.

◊   The sump pump is in (G/A/F/P)_____ condition. The pump did (y/n)_____ operate properly when tested at the time of the inspection. The sump pump is (y/n)_____ plugged into an outlet with a GFCI for safety. The sump drainage line does (y/n)_____ have a required backflow preventer. This is a check valve inside one section of the line to help prevent the drainage water from flowing backwards towards the sump pit after it is pumped out. The sump drainage line does (y/n)_____ discharge the water at least 5 feet away from the exterior foundation of the house. This is required to prevent the water from flowing back into the sump pit after it has already been pumped out from the lower level of the house.

◊   It is *highly* recommended that you check the local building department to determine if the subject property is located in a designated flood hazard zone. A *"flood hazard zone"* is an area where the government has determined that there is a potential of the area becoming flooded from time to time. Flood maps are located in every town hall and are available to the public to view. If a house is located in a flood hazard zone, then the homeowner should obtain flood hazard insurance in addition to normal homeowner and title insurance policies.

_____
_____
_____
_____
_____
_____
_____
_____
_____
_____
_____

Additional Comments

The Interior Home Inspection #1

Kitchen

◊ The kitchen walls and floors are in (G/A/F/P)_____ condition. There are (y/n) _____ signs of structural problems or abnormal settlement cracks that need further evaluation by a licensed contractor.

◊ The kitchen floor covering is in (G/A/F/P)_____ condition. The floor covering is made of _____.

◊ The kitchen cabinets are in (G/A/F/P)_____ condition. The cabinets are (y/n)_____ securely fastened to the walls and/or floor. It is recommended that the homeowner install child guards on the cabinets and drawers if there are any children in the home.

◊ The kitchen countertop is in (G/A/F/P)_____ condition. The countertop is (y/n)_____ securely fastened.

◊ The kitchen faucet was tested and there was (y/n)_____ adequate hot water. There were (y/n)_____ leaks in the faucet and/or underneath the sink. There is (y/n)_____ a properly operating spray attachment in the sink area.

◊ There was (y/n)_____ water filter device connected to the kitchen sink water lines. A water filter device is *highly* recommended for health reasons.

◊ There are (y/n)_____ an adequate number of outlets for modern usage in the kitchen. The outlets do (y/n)_____ have the required three prongs with GFCI protection for safety.

◊ The Owner or Realtor stated that the appliances are (y/n)_____ being sold with the house. The appliances were (y/n)_____ spot checked for proper operation. The inspector *cannot* properly evaluate appliances during a limited time inspection. All owners manuals and operating instructions should be obtained from the Owner.

_____
_____
_____
_____
_____

Bathrooms

◊   There is (#)_____ bathroom(s) in the house.

◊   The bathroom walls and floors are in (G/A/F/P)_____ condition. There are (y/n)_____ signs of structural problems or abnormal settlement cracks that need further evaluation by a licensed contractor.

◊   The bathroom floor covering is in (G/A/F/P)_____ condition. The floor covering is made of _____.

◊   The bath tub and shower area did (y/n)_____ have loose or cracked sections. This area was (y/n)_____ in need of grout/caulk at the time of the inspection. Grout is needed to prevent water leaks behind the walls and floors.

◊   The bathroom cabinets are in (G/A/F/P)_____ condition. The cabinets are (y/n)_____ securely fastened to the walls and/or floor. It is recommended that the homeowner install child guards on the cabinets and drawers if there are any children in the home.

◊   The bathroom sink top(s) is in (G/A/F/P)_____ condition. The sink top(s) is (y/n)_____ securely fastened.

◊   The bathroom sink, tub and/or shower faucets were tested and there was (y/n)_____ adequate hot water. There were (y/n)_____ leaks in the faucets and/or underneath the sink(s). The drain stop mechanism(s) in the sink(s) and bath tub(s) are (y/n)_____ working properly.

◊   The water pressure and drainage noted during the testing was (y/n)_____ normal and adequate with no problems indicated. The client was (y/n)_____ present during the testing.

◊   There are (y/n)_____ an adequate number of outlets for modern usage in the bathroom(s). The outlets do (y/n)_____ have the required three prongs with GFCI protection for safety.

◊   Any Jacuzzi or hot tub devices *cannot* be evaluated during a limited time inspection. All owners manuals and operating instructions should be obtained from the Owner.

_____

_____

_____

_____

_____

_____

_____

_____

Floors and Stairs

◊ The floors are (y/n)_____ in sound structural condition. There are (y/n)_____ signs of sagging or uneven sections due to abnormal structural settlement.

◊ There are (y/n)_____ finished hardwood floors noted in the rooms. The hardwood floors are in (G/A/F/P)_____ condition where visible.

◊ There are (y/n)_____ carpets noted in the rooms. The carpets are in (G/A/F/P)_____ condition. The inspector was (y/n)_____ able to check underneath the corner of some of the carpeting and/or in the closets to determine what type of flooring is underneath. _____ flooring was noted underneath the carpets and/or in the closets.

◊ If the Owner has any pets it is recommended that all carpets be fumigated or removed prior to taking possession of the house. This will prevent the possibility of moving-in and finding fleas in the carpeting.

◊ All staircases are (y/n)_____ sturdy and the steps do (y/n)_____ have even and safe stair heights to help prevent tripping hazards. All stairs over two steps in height do (y/n)_____ have secure handrails for safety. All handrails do (y/n)_____ have closely spaced posts to help prevent children from falling through the railings. There are (y/n)_____ light fixtures and light switches at the top and bottom of all stairways for safety.

◊ All windows at the base of staircases *must* have a sill height of at least 36 inches above the floor. This will help prevent someone from falling through the window in the event of a fall down the stairs. If the sill is less than 36 inches high, a window guard **must** be installed as a precautionary measure.

_____
_____
_____
_____

Walls and Ceilings

◊ The walls and ceilings are (y/n)_____ in sound structural condition. There are (y/n)_____ signs of abnormal settlement cracks or structural problems that need further evaluation by a licensed contractor. There are (y/n)_____ minor settlement cracks noted in the walls and/or ceilings that appear to be from the normal expansion and contraction of the construction materials.

◊ The interior walls are made of (sheetrock/plaster/etc)_____.

◊ There are (y/n)_____ signs of water stains or water damage in the visible areas of the walls and/or ceilings. Any water stained areas will have more extensive damage behind the walls and ceilings that is not visible to the inspector and *must* be evaluated by a reputable contractor. The finished coverings will have to be removed to expose this hidden area in order to view the damage.

◊   The interior walls and ceilings do (y/n)_____ need to be painted at this time. All settlement cracks and damaged areas should be patched and sealed over the next time the interior is re-painted. The inspector *cannot* determine if there is any lead paint in the house. This can only be determined by a reputable laboratory analysis which should be performed prior to closing if any doubts exist.

◊   There is (y/n)_____ wallpaper noted on the interior walls. The wallpaper is in (G/A/F/P)_____ condition and there were (y/n)_____ signs of peeling or aging sections. Removing any existing wallpaper is a time consuming job that can be expensive. All cost estimates should be obtained. You have to be careful removing wallpaper from sheetrock walls because you can pull the cardboard paper covering off the sheetrock walls along with the wallpaper.

◊   There are (y/n)_____ smoke detectors noted on all levels in the home. The smoke detectors were (y/n)_____ able to be spot checked for proper operation. The smoke detectors do (y/n)_____ appear to be in operating order. The smoke detectors are operated by (batteries/electricity)_____. Smoke detectors are **required** on all levels of a residence and heat detectors are recommended in the garage, attic and lower level areas. If the smoke detectors are battery operated it is *highly* recommended hat you replace all batteries after moving in. If the smoke detectors have a *"hard wired"* installation, obtain all operating instructions from the homeowner. Smoke detectors **cannot** be properly evaluated during a limited time home inspection.

_____

_____

_____

_____

Windows and Doors

◊   The windows are in (G/A/F/P)_____ condition and are (y/n)_____ operating properly. The doors are in (G/A/F/P)_____ condition and are (y/n)_____ operating properly. The windows and doors were spot checked by opening and closing at least one window and door in each room.

◊   There were (y/n)_____ cracked or broken panes of glass noted at the time of the inspection. There were (y/n)_____ broken vacuum seals noted in the thermal windows and/or doors. When this seal is broken, it allows moisture to get trapped between the two window panes of thermal glass and leaves condensation stains. Broken vacuum seals can sometimes be repaired by a glass service company but estimates should be obtained prior to closing.

◊   The window locks are (y/n)_____ operating properly. The door locks are (y/n)_____ operating properly. The locks were spot checked by the inspector. There are (y/n)_____ double key locks noted on the doors. *"Double key"* locks require a key to exit and enter through the door. The purpose of these locks is so that if a burglar breaks a door window, they cannot just turn a bolt and open the door, instead they will need a key to open the lock. **However, in some areas these locks are against the local fire codes because a key is needed to exit in the event of an emergency which can cause people to get trapped inside the house during a fire.** Check with the local fire and building departments for their recommendations about door locks. All locks *must* be replaced by a licensed locksmith upon taking possession of the house for security reasons.

Fireplaces

◊ There is (#)_____ fireplace(s) in the house.

◊ The fireplace(s) has (y/n)_____ signs of structural problems. The mortar joints and/or firebox area are in (G/A/F/P)_____ condition.

◊ The fireplace(s) has (y/n)_____ signs of back-smoking problems. Back-smoking is caused by downdrafts in the chimney flue which cause the smoke to come back into the house. Signs of back-smoking are black deposits, called *"creosote,"* on the front of the fireplace and mantel.

◊ The fireplace(s) does (y/n)_____ have a properly operating sliding screen cover and glass doors for safe and energy efficient operation of the fireplace(s).

◊ The fireplace damper(s) is in (G/A/F/P)_____ condition and is (y/n)_____ operating properly. The *"damper"* is the metal door inside the top of the firebox area which is opened while a fire is burning and closed when the fireplace is not in use. If the damper does not open and close properly or is very rusty it must be replaced.

◊ The inspector was (y/n)_____ able to view inside the chimney flue(s). The flue lining(s) is made of (brick/tile/cement/etc)_____. The visible areas of the flue lining(s) are in (G/A/F/P)_____ condition. There are (y/n)_____ signs of an excessive buildup of creosote inside the flue lining(s). All chimney flues need to be swept and repointed by a reputable chimney sweep periodically to help prevent chimney fires from thick creosote deposits.

Attic Inspection

◊   The attic space was (y/n)_____ accessible for the inspector to make evaluations of the areas above the top floor finished ceilings.

◊   The access panel leading to the attic is located _____. There is (y/n)_____ a stairway to provide easy access to the attic. The stairway was in (G/A/F/P)_____ condition and did (y/n)_____ have sturdy handrails and steps for safe passage.

◊   It is *highly* recommended that a handrail be installed inside the attic area surrounding the access opening. This will help prevent anyone walking in the attic area from falling through the access opening. Do not wait for an accident to happen, install a handrail now!!

◊   The attic floor is (y/n)_____ covered so that this area can be used to store **lightweight** household items. Heavy objects are not recommended for storage in the attic area due to the excessive weight they exert on the ceiling below.

◊   The roof ridge beam appeared to be in (G/A/F/P)_____ condition where visible and accessible. The roof *"ridge beam"* is the main girder type beam at the top of the crest of the roof.

◊   The roof rafters appeared to be in (G/A/F/P)_____ condition where visible and accessible. The roof *"rafters"* are the floor joist type beams leading from the attic floor up to the roof ridge beam.

◊   The roof sheathing appeared to be in (G/A/F/P)_____ condition where visible and accessible. The roof *"sheathing"* is the sub-flooring type wood that the roof shingles rest upon. The sheathing is made of (plywood/etc)_____.

◊   There were (y/n)_____ water stains that appeared to be due to water leaks or abnormal humidity in the attic area. Often there are old water stains from prior roof leaks that have been repaired. There are (y/n)_____ bowed and/or damaged sections of the wood members in the attic roof area. Any water leaks, abnormal humidity, bowing and/or damaged wood members in the attic roof area will require repairs by a licensed roofing contractor.

◊   There are (y/n)_____ collar beams in the attic. *"Collar beams"* are generally constructed of 2 x 4 inch wood boards that are located several feet below the ridge beam. The purpose of collar beams is to *"tie"* both sides of the roof together so that all of the weight of the roof does not rest upon the ridge beam alone. Collar beams give the roof additional support and it is recommended that they be installed when not noted.

_____

_____

_____

_____

_____

Attic Ventilation

◊ It is very important that the attic be properly ventilated to help prevent excessive humidity or heat in this area. Even in cold winter months humidity can cause problems in attics and needs to be ventilated to the exterior. In the summer months attics can reach 150 degrees Fahrenheit, which adds a big heat load on the house.

◊ There size of the attic vent(s) do (y/n)_____ appear to be adequate to provide proper ventilation in this area.

◊ The screens on the attic vents are in (G/A/F/P)_____ condition. The screens need to be kept clean at all times. Screens help to keep birds and bees out of the attic.

◊ There were (y/n)_____ bathroom fans noted to be discharging in the attic area. All bathroom vents and fans *must* discharge to the exterior of the house. If they discharge in the attic they will create moisture problems.

◊ There are (y/n)_____ soffit vents. *"Soffit vents"* are vents at the base of the roof where it overhangs the exterior siding. Soffit vents are recommended to allow air to come into the base of the roof in the attic area and carry any unwanted heat or moisture out the attic gable or roof vents.

◊ There is (y/n)_____ a ridge vent. A *"ridge vent"* is a vent at the very top of the roof, above the ridge beam. Ridge vents are recommended to allow unwanted heat and moisture to escape from the attic through the top of the roof.

◊ There is (y/n)_____ a thermostatically operated power ventilator in the attic. A *"thermostatically operated power ventilator"* is a fan in the roof that operates by a thermostat. When the temperature in the attic reaches a pre-set level, the fan will automatically turn on to cool this area. When the temperature drops low enough, the fan will automatically turn off by itself.

◊ There is (y/n)_____ an attic and/or whole house fan installed in the house. Attic and house fans need to have adequate vents to discharge the air so the fan will operate properly and will not break down prematurely.

_____

_____

_____

_____

_____

_____

_____

_____

_____

Attic Insulation

◊ There is (y/n)_____ insulation noted in the floor joists of the attic area where visible and accessible. Insulation in the **attic floor joists** is required for energy efficiency in a house. Insulation is *not* needed in between the **roof rafters** because once heat has escaped through the upper level ceiling it is lost anyway. There is no sense trying to trap it in the attic, you will only be trapping unwanted moisture along with the heat in the attic if you install insulation in the roof rafters.

◊ The insulation is approximately (#)_____ inches thick. It should be at least 8 inches thick for energy efficiency. Installing additional insulation will increase energy efficiency in the house. Any air-conditioning or heating ducts in the attic must be insulated either on the exterior or the interior for energy efficiency.

◊ The type is insulation noted is made of _____. The vapor barrier is (y/n)_____ installed properly. The *"vapor barrier"* is the aluminum foil layer on one side of the insulation roll. It must always be touching the **heated** side of the building. For example, if it is installed in an attic the vapor barrier must face <u>downwards</u>. If it is installed in an unheated basement or crawl space the vapor barrier must face <u>upwards</u>. The reason for the vapor barrier is to prevent any moisture from getting trapped in between the insulation and condensing into water, which will decrease the energy efficiency. It prevents moisture problems by reflecting the heated air, which has moisture in it, back towards the heated portion of the house. If an existing layer of insulation has a vapor barrier, then if you add more insulation on top of the existing layer, the new layer added should **not** have a vapor barrier since the aluminum foil barrier will trap moisture in between both layers of insulation.

◊ The Owner or Realtor stated that they have (y/n)_____ installed insulation into areas of the structure. They stated that they do (y/n)_____ know if any prior owner's have installed insulation in the structure. Any *"blown-in"* type of insulation can have substances in it that are a health hazard. In the past, some houses had UFFI insulation blown into the walls and floors. UFFI stands for *Urea Formaldehyde Foam Insulation* and the Environmental Protection Agency, or *EPA*, has issued warnings about this type of insulation. If there is any UFFI or other type of unknown insulation in the house, it is **highly** recommended that an air sample be taken by a licensed laboratory to see if there are any health concerns with this insulation in the house.

_____
_____
_____
_____
_____
_____
_____
_____

Asbestos Insulation

◊ Many older houses will have asbestos insulation on the heating pipes. Sometimes the old cast iron boilers have asbestos on the interior insulating walls also. Believe it or not, when asbestos first came out it was <u>required</u> to be installed in all new construction. That is why so many buildings have it. It was considered a "miracle product" when it first came out because Asbestos has great insulating and fireproofing qualities. The only problem was that they did not know about the health problems associated with it until it was too late.

◊ Asbestos causes lung cancer when it comes loose from the pipes and the fibers get into the air. Asbestos fibers are like tiny daggers and when you breathe them in, they stick into your lungs and stay there. The fibers cling to dust and can be stirred up off of the floor when someone walks in a basement. There are about 5 different diseases that are related to exposure to asbestos.

◊ The Environmental Protection Agency, *(EPA),* has offices in every State which will provide anyone with free information and brochures on Asbestos, Radon Gas, Fuel Oil Leaks, Water Quality, and a lot of other environmental and health concerns that the homeowner needs to know about. Call your State EPA office and obtain their brochures for more information and advice.

◊ There is (y/n)_____ evidence of possible Asbestos insulation in the house. Asbestos insulation usually has a white color and appears to have layers of ribbed cardboard in the middle sections. It is usually wrapped in an off-white canvas covering. However, the only way to know for sure if any insulation is Asbestos is to have a laboratory take an air sample. **It is recommended that you contact a licensed Asbestos contractor or laboratory whenever there is <u>any</u> evidence of possible Asbestos insulation or if any doubts exist.**

◊ There are (y/n)_____ signs of residue or loose sections of Asbestos insulation in the house. The Environmental Protection Agency recommends that Asbestos insulation be <u>PROFESSIONALLY</u> sealed or removed from the residence by an EPA *licensed* Asbestos contractor. This means, that the homeowner, the plumber, heating contractor, or any other handyman **<u>SHOULD NOT TOUCH</u>** any Asbestos in the house. Residue from Asbestos insulation on the heating pipes is sometimes noted by the existence of small white particles on sections of the pipes. **This indicates that an Non-EPA licensed person removed the Asbestos!! It is <u>highly</u> recommended that an EPA licensed Asbestos contractor or laboratory be contacted for further evaluations prior to closing on the property if any evidence or doubts exist.**

◊ When an EPA licensed contractor removes Asbestos, they seal off the entire area where it is located and work with completely sealed suits over their bodies. They then set up a vacuum to remove **all** of the dust from the area. When the Asbestos is totally removed from the house, they then take an air sample to make sure they have not left any fibers lying around to be stirred up and breathed in later. Generally, any Asbestos behind the walls is left alone. If there is no access to it and it cannot be disturbed, there usually is not much of a health concern.

_____

_____

_____

_____

_____

Radon Gas

◊   Radon is a radiation gas that is released naturally by rocks and soil in the earth. It gradually seeps up from the ground, and as long as it goes out into the open air it is not a problem. However, if the radon seeps through cracks in the foundation floor and walls it will become trapped in the house and the radon levels will rise. As with asbestos and other environmental and health concerns, call your State Environmental Protection Agency office for their information, brochures and advice. **EPA considers radon to be the number 2 leading cause of lung cancer behind smoking, so it is not something to take too lightly.**

◊   EPA uses a reading of 4 Pico Curies per liter to determine the maximum radon level in a house before mitigation is recommended. Just to give you an idea of how Pico Curies are measured, EPA says that 1 Pico Curie is the average indoor radon level and it is equal to getting about 100 chest X-rays per year. Now that may seem very high, but to put it in the proper perspective, EPA also says that the amount of radiation you receive from a normal chest X-ray, usually is not as high as most people think. For example, with a reading of 1 Pico Curies per liter, EPA estimates that 3-13 people out of 1,000 will die from lung cancer. This is similar to a non-smokers risk of dying from lung cancer. With a reading of 4 Pico Curies per liter, it is estimated that 13-50 people out of 1,000 will die from lung cancer. This is similar to 5 times the non-smokers risk of dying from lung cancer. The lung cancer risk increases as the radon levels and time of radon exposure increases.

◊   *"Mitigation"* is the term used to reduce the problem by lowering the radon levels. When a house is mitigated, the radon contractor will seal all open cracks in the lower level walls and floors. They then drill a hole in the foundation floor, which will look like a sump pump pit. Instead of installing a sump pump in this pit, they install a fan with pipes leading to the outside of the house. In some areas, the local codes require that these pipes discharge **above** the roof line to prevent the radon from entering back into any open windows on the side of the house. The purpose of the mitigation treatment is to vent the radon gas that builds up underneath the foundation, to the exterior of the house.

◊   In some areas the radon levels tend to be higher than in other areas but **ALL HOUSES WILL HAVE SOME RADON!!** Even if it is minor trace element readings of 0.5 Pico Curies per liter, which can be a minimal health risk. However, you might not have a high radon reading today but you might have a high reading a month from now. Or you might have a high reading and your neighbor might not. The reason for this is that radon is a radiation gas that is unstable and the levels fluctuate often. There are a lot of factors that effect the radon level in a house, such as: 1) The time of the year and the climate. 2) The type of soil and rocky terrain in the area. 3) The type of construction of the building. 4) And there are other factors also. This is why EPA recommends that you re-test for radon every 6 months to make sure that the levels in your house are acceptable. Believe it or not, radon can even be found in water.

◊   Radon testing is usually done with a small, round metal canister with charcoal inside. The canister is left in the house for about 3-5 days and then it is sealed and mailed back to the radon lab for analysis. Generally, all the canister does is absorb the air in the room where it is placed so the lab can analyze the radon levels. The canister does not present a health risk to the occupants of the home. Radon testing canisters *must* be purchased from a licensed EPA laboratory with sophisticated analyzing equipment. **Do not just buy radon cans off the shelf of the local hardware store. The reason for this is that what makes a radon reading accurate is not the canister you use, but the sophistication of the labs analyzing equipment. You could send the same canister to two different labs and get two totally different radon level readings.**

◊   Most labs recommend that you place the canister about 3 feet above the floor in the lowest area of the house. The Environmental Protection Agency feels that if a basement has the potential to become a livable area in the future, then that is where the radon levels should be measured. The basement is where you are going to get the highest radon reading in the home. Any readings on the first floor will generally be lower then basement readings.

Additional Comments

The Exterior Home Inspection #1

Roof

◊   There are (y/n)_____ tree branches overhanging the roof area. There should not be any tree branches overhanging the roof, which can cause damage to the shingles.

◊   The inspector was (y/n)_____ able to view the roof adequately from the following location(s) _____. All judgments and evaluations are made from that perspective.

◊   There are (y/n)_____ bowing sections of the roof ridge beam, roof rafters and/or the roof sheathing which would indicate repairs being needed.

◊   The roof covering/shingles installed on the house is made of _____ _____. The roof covering/shingles appears to be approximately (#)_____ years old. It is in (G/A/F/P)_____ condition. There are (y/n)_____ signs of rapidly aging and damaged areas of the roof covering.

◊   The average life expectancy of this type of roof covering/shingles is (#)_____ years. The life expectancy of all roofs depends upon many factors, some of which are the quality of the roof covering, the quality of the installation, the climate and exposure to the elements and the maintenance given to the roof over the years.

◊   The slope of the roof that has a southerly or southwesterly exposure faces the sun more often and can become brittle and show signs of aging faster. The slope of the roof that has a northerly or northeastern exposure is more apt to have mold and decay fungi due to the lack of sunlight.

◊   There appears to be (#)_____ layer(s) of shingles on the roof. Some local building codes allow there to be up to 3 layers of shingles on a roof. However, this adds too much weight to the structure and it is only recommended that there be 2 layers as a maximum. Also, when shingles are placed over an existing layer they tend to have a poor cosmetic appearance and it cuts down their life expectancy. When there are 2 or more layers of roof covering installed presently, then you will have to strip these layers off during the next re-roofing. This is much more expensive than just having a new layer put over the existing shingles. *Especially* if the roof sheathing has to be replaced as well.

_____

_____

_____

_____

Chimney

◊   There is (y/n)_____ a chimney for the subject property.

◊   The chimney(s) is (y/n)_____ leaning. Any leaning conditions indicate a serious structural problem where they chimney may need to be rebuilt.

◊   The chimney(s) is in (G/A/F/P)_____ condition. There were (y/n)_____ signs of maintenance and repairs being needed due to the exposure to the elements.

◊   The mortar joints are in (G/A/F/P)_____ condition. There is (y/n)_____ evidence that they need to be re-pointed. *"Re-pointing"* refers to putting more cement in the joints mortar joints. The mortar joints between the construction materials need to be checked periodically for deterioration problems to help prevent water penetration.

◊   The chimney(s) is (y/n)_____ made of metal materials. The metal chimney materials noted are in (G/A/F/P)_____ condition. Metal chimneys need to be checked periodically for rust and water leaks.

◊   There is (y/n)_____ an antenna attached to the chimney or the roof. The connections of an antenna must be kept caulked so that water will not enter the house. If there is cable TV in the house and the antenna is no longer in use, then it should be removed. Antennas add stress to the roof and chimney when they move around in the wind, which can create water leaks.

◊   The chimney(s) lining was (y/n)_____ visible from vantage points on the subject property. There is (y/n)_____ a weather cover noted above the top of the chimney stack to prevent water from entering the flue. There is (y/n)_____ a screen noted over the top of the flue stack to prevent animals, such as squirrels, raccoons and birds, from entering the flue.

_____

_____

_____

_____

Siding

◊   The siding on a house is used to provide weather protection. The siding does not support the building structurally, unlike a load bearing wall which does give structural support to the house.

◊   The siding installed on the house is made of _____ _____. It is in (G/A/F/P)_____ condition. There are (y/n)_____ signs of rapidly aging and damaged areas of the siding.

◊   The average life expectancy of this type of siding is (#)_____ years. The life expectancy of all siding depends upon many factors, some of which are the quality of the siding, the quality of the installation, the climate and exposure to the elements and the maintenance given to the siding over the years.

◊   The siding does (y/n)_____ need to be painted or stained at this time. Painted wood will have a more uniform appearance but needs more maintenance. Stained wood will have spotty areas due to the wood absorbing the stain unevenly in some sections. However, staining wood will last a lot longer than painting wood.

◊   All joints around windows and doors do (y/n)_____ appear to be caulked properly to help prevent water penetration and increase energy efficiency.

◊   All areas of the siding are (y/n)_____ at least 8 inches above the soil all around the structure. This is required to help prevent termite and rot problems.

_____
_____
_____
_____
_____
_____
_____
_____

Fascia, Soffits and Eaves

◊   The fascia, soffits and eaves are the molding areas at the bottom of the roof and the top of the siding. It is the small area where the roof overhangs the sides of the house.

◊   The fascia, soffits and eaves are in (G/A/F/P)_____ condition. There are (y/n)_____ signs of rapidly aging and damaged areas at this time. They are (y/n)_____ in need of painting at this time.

◊   There are (y/n)_____ vents noted at the bottom of the roof overhang area. When vents are noted, it indicates that the attic area may have soffit vents. *"Soffit vents"* allow air to enter the bottom of the attic area to help to remove any unwanted heat and moisture from the house attic.

_____
_____
_____
_____

Gutters, Downspouts and Leaders

◊ **Gutters** are installed along the bottom edge of the roof to catch the rainwater running off the roof. **Downspouts** are installed near the ends of the gutters and are used to drain the water from the gutters. **Leaders** are installed at the bottom of the downspouts to direct the rainwater away from the side of the house.

◊ The gutters, downspouts and leaders are made of _____. They are in (G/A/F/P)_____ condition. There is (y/n)_____ evidence of loose or leaning sections that need to be secured. The gutters, downspouts and leaders do (y/n)_____ appear to be clogged with leaves and/or twigs at this time.

◊ There are (y/n)_____ an adequate number of gutters, downspouts and leaders on the house. There should be at least 1 downspout for every 30 feet of gutter to help prevent excessive weight damaging the gutters due to the rainwater.

◊ All downspouts do (y/n)_____ have leaders to pipe the rainwater at least 5 feet away from the foundation to help prevent water problems in the lower level areas.

◊ There are (y/n)_____ downspouts that drain directly into the ground. These generally lead to dry wells or underground drainage lines. At the time of the inspection, they did (y/n)_____ appear to be clogged. They need to be checked periodically for clogging due to leaves and small animals getting stuck in them.

_____

_____

_____

_____

Windows, Screens and Storms

◊ The exterior window frames are in (G/A/F/P)_____ condition. There is (y/n)_____ evidence of rot and/or damaged areas. They do (y/n)_____ need to be painted or stained at this time.

◊ The storm windows and/or screens over the windows are in (G/A/F/P)_____ condition. Storm windows are *highly* recommended in colder climate areas. More heat is lost in a house through the windows than through any other area. Storm or thermal windows can reduce the heat loss by as much as 50%.

_____

_____

_____

Entrances, Steps and Porches

◊ All accessible entrances, steps and porches are determined to be in (G/A/F/P)_____ condition. There is (y/n)_____ evidence of structural problems.

◊ The landing platform(s), which is the area in front of doors, does (y/n)_____ have a large enough space to safely open the door while someone is standing there. This is required to help prevent someone from being knocked down the steps when the door is opened.

◊ There are (y/n)_____ handrails for <u>all</u> stairs that are more than two steps in height. The handrails are (y/n)_____ loose and/or decayed. Handrails **must** be installed when not noted and maintained periodically to prevent any tripping hazards.

◊ The steps do (y/n)_____ have uneven and/or damaged sections that require repairs at this time. All steps *must* have an even and uniform height and be properly maintained so that there are no tripping hazards.

◊ There are (y/n)_____ wood stairs with the wood base resting directly on the soil. When wood stairs are noted, the base of the wood should be resting on concrete pads above the soil. This will help prevent rot and termite infestation.

_____

_____

_____

_____

Walks

◊ All accessible walks are determined to be in (G/A/F/P)_____ condition.

◊ There are (y/n)_____ uneven and/or damaged sections in the walks that require repairs at this time to prevent any tripping hazards. There are (y/n)_____ weeds growing in between the walkway sections that need to be removed.

◊ There is (y/n)_____ a sidewalk at the street. When a sidewalk is noted, it is recommended that you check with the local building department to determine whose responsibility it is to repair the sidewalk. In most areas, the homeowner is responsible for repairing and cleaning the sidewalk in front of their home.

_____

_____

_____

_____

Patios and Terraces

◊   There is (y/n)_____ a patio(s) associated with the subject property.

◊   The patio(s) is determined to be in (G/A/F/P)_____ condition. There are (y/n)_____ uneven and/or damaged sections in the patio(s) that require repairs at this time to prevent any tripping hazards. There are (y/n)_____ weeds growing in between the patio sections that need to be removed.

◊   The joints of the patio are (y/n)_____ properly caulked. If the patio touches the side of the foundation, then it should be well caulked and sloped away to prevent water from draining towards the house.

◊   In most areas building permits and approvals are needed to build patios. Determine if they have been obtained for any patio(s) at the site.

_____

_____

_____

_____

_____

Decks

◊   There is (y/n)_____ a deck associated with the subject property. Decks <u>always</u> require building department approvals because of the safety concern if they are improperly built. Determine from the local municipality if all valid permits and approvals have been obtained for any deck construction.

◊   There is (y/n)_____ evidence of rotted and/or damaged sections of wood that need to be replaced.

◊   The deck railings are (y/n)_____ sturdy. The deck railings are (y/n)_____ properly spaced for safety. The deck perimeter railings *must* be spaced so that a maximum gap of 4 inches exists between them. This is to help prevent small children or dogs from falling through the openings.

◊   The main beam is (y/n)_____ lag bolted to the side of the house. There *must* be lag bolts in the main beam, called the *"header beam,"* where the deck is attached to the side of the house. Lag bolts are a **far superior** way to support the deck, as opposed to just using nails.

◊   The floor joists of the deck do (y/n)_____ have steel support hangers. The floor joists of the deck *must* have steel support hangers to give them additional support, as opposed to just nailing them.

◊   The deck support posts and girders do (y/n)_____ have steel support brackets. All deck support posts and girders *must* have steel support brackets at the base **and** at the top for support. The base of the support posts are (y/n)_____ resting on concrete pads. The base of the posts should be resting on a concrete pads with steel brackets to keep the wood from being in contact with the soil.

The Exterior Home Inspection #1   www.nemmar.com

_____
_____
_____
_____
_____
_____
_____
_____

Walls and Fences

◊   There is (y/n)_____ a wall(s) and (y/n)_____ a fence(s) associated with the subject property.

◊   Retaining walls are used to support the soil in areas that are dug into the earth, such as driveways or yards. The retaining wall(s) noted is made of _____. The wall(s) is in (G/A/F/P)_____ condition. The retaining wall(s) is (y/n)_____ leaning at this time. Any leaning conditions indicate that repairs **must** be made to prevent the wall from moving any further or collapsing.

◊   The retaining wall(s) does (y/n)_____ appear to have adequate weep holes at the base. The purpose of *"weep holes"* is to relieve the pressure by allowing any water that builds up behind the wall to drain safely away.

_____
_____
_____

◊   The fence(s) is made of _____. The fence(s) is in (G/A/F/P)_____ condition. It is *highly* recommended that you check with town hall to determine if the fence is located within the subject property line. Often the homeowner or a neighbor will have a fence installed and the contractor will just guess where the property line is. This will lead to a property line encroachment.

_____
_____
_____

Drainage and Grading

◊   The soil grading next to the foundation is (y/n)_____ properly sloped away from the house. The soil must slope away from the structure to help prevent the rainwater from building up next to the foundation. In most cases, the soil only needs to slope about 1/2 inch for every foot away from the house to properly drain the water.

◊   The bushes, shrubs and/or trees are (y/n)_____ properly pruned away from the side of the house. The bushes, shrubs and trees must be pruned away from the side of the house to allow enough sunlight and air next to the foundation to help prevent rot and wood destroying insect problems.

_____
_____
_____
_____

Driveways

◊   There is (y/n)_____ a driveway on the site. Driveways that do not have a finished surface and are made of gravel and dirt are not recommended. Often they have holes which are a tripping hazard which need to be filled. Unfinished driveways can lead to people tracking dirt into the house and they cannot be shoveled for snow removal in colder areas

◊   The driveway is made of (dirt/gravel/asphalt/concrete/etc) _____
The driveway does (y/n)_____ have holes, cracked and/or uneven sections that need to be repaired. Asphalt driveways need to be sealed with a driveway sealer every 2 or 3 years to prevent them from drying out and cracking. Concrete driveways need to be patched periodically. Gravel driveways get potholes that need filling.

◊   There is (y/n)_____ a water removal drain at the base of the driveway. These drains need to be checked periodically due to problems with clogging.

_____
_____
_____
_____

Garage

◊ There is a (y/n) or (built-in/attached/detached)_____ garage on the premises. The garage has a capacity for (#)_____ cars. The benefit of having an attached or built-in garage is that you can park the car and enter the house without worrying about the weather conditions. A detached garage is safer in the event that a car is left running by mistake, or if there is a garage fire.

◊ The garage was in overall (G/A/F/P)_____ condition. The garage doors did (y/n)_____ come down with excessive force when checked. The door springs will need to be adjusted and lubricated periodically. Doors that come down with excessive force can crush a child if they are caught underneath.

◊ The electric door openers did (y/n)_____ operate properly. The automatic reverse function was tested and did (y/n)_____ operate properly. Electric door openers *must* have an *"automatic reverse"* function that is working properly. This is a setting in the opener that will stop or reverse the direction of the doors if a person or a car gets caught underneath. It must be checked periodically for safety.

◊ All visible ducts and water pipes in the garage were (y/n)_____ insulated. All ducts and water pipes in the garage need to be insulated to prevent them from freezing and for maximum energy efficiency.

◊ There are (y/n)_____ excessive gas and/or oil drippings on the garage floor. Any oil or gas drippings *must* be cleaned to help prevent any fires.

◊ The garage walls and ceilings do (y/n)_____ appear to have an adequate fire resistant covering. The garage walls and ceilings *must* be covered with fireproof sheetrock to help prevent the spread of any fires. Masonry walls and ceilings are an acceptable fireproof covering.

◊ The garage does (y/n)_____ have a fire resistant entry door leading to the house. The door does (y/n)_____ have a properly operating self-closing device. If the garage is attached or built-in there *must* be a fireproof entry door leading to the house. Also, this door *must* have a self-closing device to prevent the door from being left open so car exhaust fumes or fires cannot spread easily into the house.

_____
_____
_____
_____

Other Exterior Structures

◊ There is (y/n)_____ an exterior shed or other structure on the property.

_____
_____
_____

Swimming Pools

◊ There is an (y/n) or (above-ground/in-ground)_____ swimming pool on the property. **All** swimming pools require local town permits and approvals which need to be verified.

◊ There is (y/n)_____ a fence surrounding the pool. **All** swimming pools need to have fences surrounding them to prevent any children from falling into the water and drowning. Special homeowners insurance is needed with swimming pools.

◊ There are (y/n)_____ tripping hazards noted around the pool area which need to be repaired.

◊ The pool walls did (y/n)_____ appear to have evidence of leaks, cracks and/or bulging sections that need to be evaluated by a reputable pool contractor.

_____

_____

_____

_____

Wood Destroying Insects

◊ There is (y/n)_____ evidence of wood destroying insect damage in the visible and accessible areas of the subject property. If there was any aspect of performing a home inspection that the inspector is believed by some clients to have X-ray vision, then this one takes the prize! If *Superman* really existed, he would make a fortune as a termite inspector! Many home inspectors get complaints from former clients because they did not notice termites that were behind the finished and covered walls and/or floors. ***THE INSPECTOR CANNOT BE HELD RESPONSIBLE FOR AREAS AND ITEMS THAT ARE NOT ACCESSIBLE OR NOT VISIBLE!!***

◊ There are many different types of wood destroying insects, including 70 species of termites throughout the world. The wood destroying insects that generally concerns people the most are: Subterranean Termites, Dry Wood Termites, Damp Wood Termites, Powder Post Beetles, Carpenter Ants, and Carpenter Bees.

- o *Termites* eat the wood and turn it into food. They have one celled organisms in their digestive tracts which converts the cellulose of wood back into sugar which they can digest. In forests termites are beneficial in the fact that they help to decompose fallen trees and stumps and return the wood substances to the soil to be used again by other trees. Termite damaged wood will have channels in it and there will not be any sawdust around.

- o *Powder Post Beetle* larvae eat the wood and lay their eggs in it. They <u>cannot</u> convert the cellulose in the wood to sugar and therefore, must get their nourishment from the starch and sugar which the tree had stored in the wood cells. To these insects the cellulose in the wood has no food value and is thus ejected from their bodies as wood powder or *"frass."* They derive nourishment from the starch and sugar in the wood.

- *Powder Post Beetle* damaged wood will crumble like sawdust when you probe it. A common indication of these insects is the existence of tiny holes in the wood. If only a single generation of this beetle larvae has fed within some wood, it is usually still structurally sound. But the feeding of generation after generation is what reduces the interior of the wood to a mass of powder. Before the female will attach her eggs to a piece of wood, she first actually tastes the wood to be sure it contains enough sugar and starch to nourish her offspring. If she is prevented from doing this due to any covering on the wood, such as paint, varnish, stain, etc, she will not deposit her eggs in the wood and it will not be reinfested with another generation of Powder Post Beetle larvae. That is why there should not be any untreated wood around the house.

- *Carpenter Ants* and *Carpenter Bees* merely excavate the wood to make nests. The damage they cause will leave sawdust outside the wood channels.

◊ An important fact to remember when getting a corrective wood destroying insect treatment on a house is that it is recommended that <u>all</u> of the damaged wood be removed and replaced. This will ensure that any damaged areas of wood are re-supported with good, solid lumber. Another reason for this is that there is *no way* to tell down the road if the wood had gotten the termite damage before or after the corrective treatment was performed.

◊ They say there are 2 kinds of houses: **Houses that <u>have</u> termites; and Houses that <u>will have</u> termites.** That's a fact. All houses will get termite damage to some extent eventually. Sometimes builders will install a termite shield along the top of the foundation wall. A *"termite shield"* is a small metal guard or molding placed at the top of the foundation walls. It is similar in purpose to installing a cap plate at the top of concrete block walls. However, these shields **do not** prevent termites. The only benefit from them is that they *might* deter termites or make it a little more difficult for them to reach the wood.

◊ There are many ways to help prevent wood destroying insect and rot damage:

- Use pressure treated lumber whenever replacing or constructing anything on the site or in the house. Pressure treated lumber has a greenish color to it. It is rot and termite resistant for up to 40 years. The most common type of pressure treated wood is called *CCA 40*. There is also a *CCA 60* pressure treated lumber that has a higher pressurization and life expectancy than CCA 40 does.

- The way the pressure treated process works is they will take the lumber and place it in large vats of chemicals where it will sit until the chemicals are absorbed sufficiently into the wood. The chemicals they use are *copper, chromate and arsenic*. The arsenic deters any wood destroying insects. The type of wood that is used for pressure treated lumber is *Southern Yellow Pine*. The reason for this is that it is the best lumber to use for the chemical process performed.

- Another way to help prevent wood destroying insect and rot damage is to keep all wood siding and trim work about 8 inches above the soil to make it more difficult for termites to get to their food source. Keep all bushes and shrubs pruned and keep the soil and drainage leaders sloped away from the foundation. This will help prevent any dark and moist areas that attracts termites.

- Get a *preventive* wood destroying insect treatment before finding any damage, as opposed to a *corrective* treatment after the damage is found. Preventive treatments are usually about **half** the cost of a corrective treatment because they only treat around the exterior perimeter of the house.

◊ It is *highly* recommended that you get a corrective treatment if the inspector finds damage and a preventive treatment if the inspector does not find any damage. Since all houses do get some form of wood destroying insect damage over time, you might as well eliminate the problem ahead of time when it is less expensive to do so. It is also easier to sell a house with a preventive treatment before any damage is found, as opposed to a corrective treatment done after there was damage found.

◊ There are certain houses that many Pest Control Operators, *(PCO)*, **will not** treat, or else there will be only a limited number of them that will treat the house. Some of these houses that can be difficult to treat are:

- Houses with on-site well water systems. The PCO has to worry about contaminating the well water supply. If the well is less than 100 feet from the house your chances of finding a PCO to treat diminish even further.

- Houses that have brick foundation walls. The PCO has to worry about contaminating the house by seepage through the brick walls.

- Houses that have air ducts embedded in the lower level cement floor for the heating or air-conditioning systems. The PCO has to worry about contaminating the air ducts.

- If the inspection is being conducted on a condominium then the By-Laws or Prospectus of the Condo/Owner's Association may have requirements that can in some way restrict wood destroying insect treatments.

Additional Comments

Safety Concerns #1

Safety Concerns

◊ Items such as tripping hazards in the steps, walks and patios, loose and missing handrails, proper deck construction and guardrails, leaning retaining walls, and loose electrical grounding cables, can sometimes seem like minor items to repair. However, these are things that can cause someone to get **seriously** hurt if they are not repaired immediately and properly.

◊ An uneven section in a walkway might not seem like much but what happens if the person that falls hits their head. A leaning retaining wall will crush a child if it falls on top of them! A missing or loose handrail could cause someone to fall down the steps. **The point we want to make very clear is don't tank chances with safety items!! You could end up costing someone their *LIFE!!* Repair all hazardous conditions immediately!**

Additional Comments

Home Inspection Conclusion #1

Conclusion

◊ Many times the Realtor, the seller or some third party to a transaction that is involved with the deal, will tell the inspection client that something, such as, the roof, the heating system, etc., was just recently replaced. Or many times they will say that all the building permits and approvals have been obtained for an addition, deck, or some other aspect that was a change to the original construction of the building or site. If this is the case with the subject property, then it is *highly* recommended that you obtain all receipts and documentation for the work performed and that you check with the local building department to make sure that this information is accurate!! Whenever you upgrade the roof, heating system, air-conditioning system, electrical system, etc.; put an addition on a house; add a deck; install a swimming pool; or make any changes to a house or a site from the original construction, you have to file the necessary permits with the local municipality. The reason for this is that the local building department inspectors have to check the work to make sure it meets all the necessary building codes in that town.

◊ **Do not** just take it for granted that the permits and approvals have been obtained for any work performed!! Many people will upgrade from the original construction without filing for permits. They might do the work themselves or else they hire a contractor who does not know what he is doing and he will not file any permits for the work performed. You should go down to town hall *personally* to check **all** records pertaining to the subject property!! This will enable you to verify all information in the real estate listing and what has been told to you about the subject property. If you send a Realtor or another third party to town hall to check the records, and they miss something, it is **YOU** that is going to have to deal with the problem later!! This will end up costing you time and money. So you should go and check it yourself, as opposed to just sending someone else to do it for you. At town hall the records will show the amount of taxes on the house, if there are any building violations, any easements, encroachments or problems with the title and deed of the property, and a lot more. All of this information is very valuable to you and many people do not even realize how much information is open for the public to view at their local town hall records department.

◊ The decision to buy or not to buy, and what repairs are done to the subject property is totally up to you. A home inspector can inform you of the current condition of the accessible and visible areas of the subject property only! A home inspector *is not* an appraiser determining **market value** of the subject property. A home inspector only determines the **condition** of the subject property. There is a big difference between the two. The point is that only a very well trained and qualified Real Estate Appraiser can determine market value - not a home inspector.

◊ Any aspects of concern brought forth in the on-site inspection or in the written report *must* be checked out by a reputable, licensed contractor if any doubts exist. You are encouraged to call a contractor and obtain written estimates, on their own, for any areas of concern or repairs needed.

**This written report is not assignable to any third parties in any way, shape or form.
No part of this work shall be reproduced, stored in a retrieval system, or transmitted by any means, electronic, mechanical, photocopying, recording or otherwise.
This report format is © Copyright 1992-2004 www.nemmar.com . All rights are reserved.**

Home Inspection Photo Pages

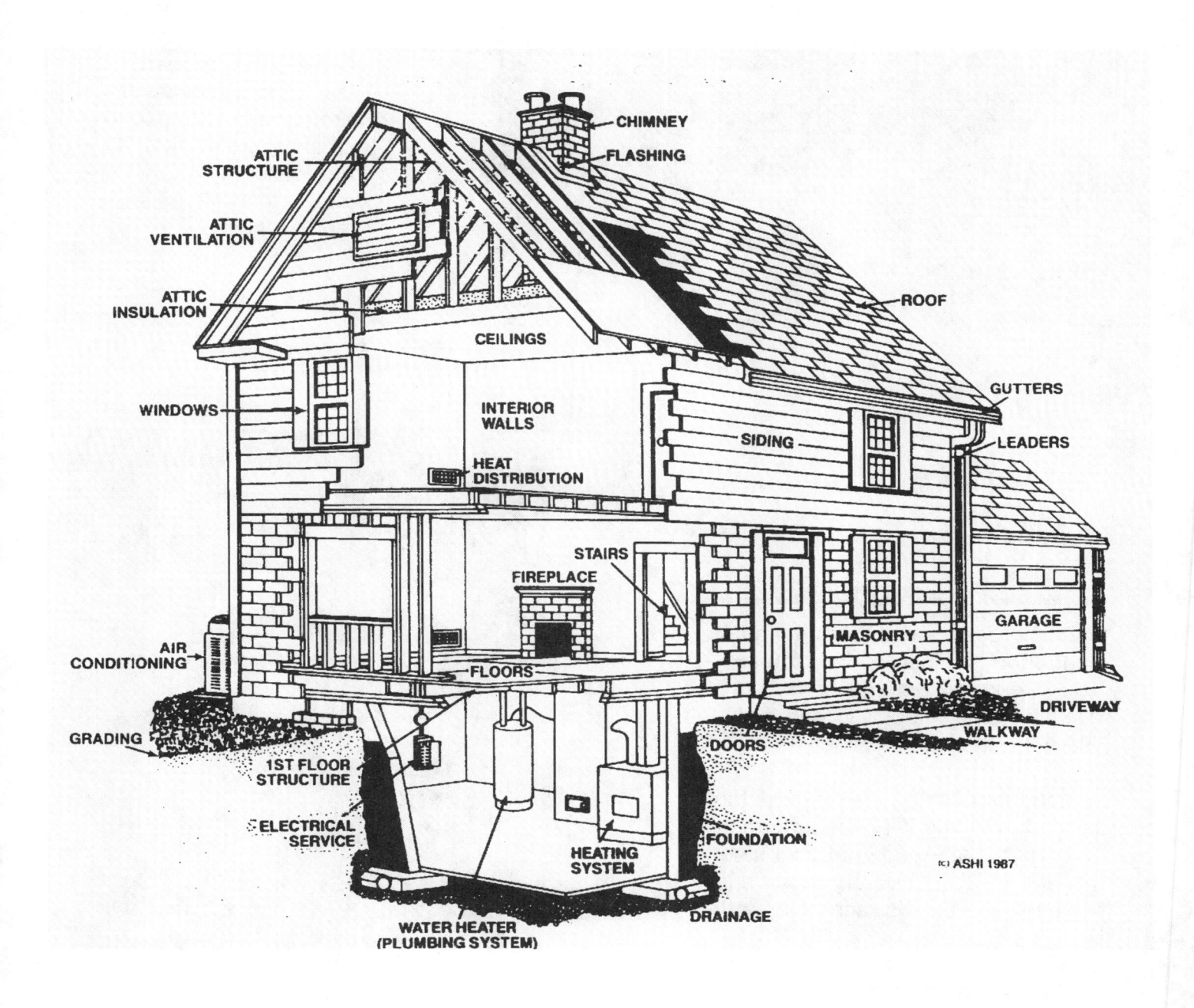

P 1. *No, that's not a submarine periscope!* It's the oil supply pipe that is used to fill this underground oil tank. The longer, iron pipe by the house, is the tank vent pipe. This allows air to escape so the tank can be filled properly. Underground oil tanks are a major expense to clean up. EPA laws have become much more strict with oil and lead problems.

P 2. There is a small fuel level gauge on the top of this interior oil tank. This interior oil tank has some signs of oil stains on the top. This can happen when the tank is overfilled.

Notice the patch on the bottom of this tank. Due to humidity, the bottom of these tanks often rust out over time. The patch is a temporary repair and replacing the tank is recommended.

The oil supply line has a firematic shut-off valve for safety.

The copper oil supply line is embedded in the concrete floor to protect it against damage.

P 3. This is a properly installed water main line. The electrical grounding cable is clamped on both sides of the meter.
The bell shaped water pressure reducing valve indicates there is strong water pressure from the street main water line.
There are shut-off valves on each side of the meter for easy replacement of the meter. The lever shut-off valve *(upper right)* is more reliable than the knob type valve.
water meter

P 4. This is called a disaster!! This water main has so many problems that I don't know where to begin.
The water meter is very old, outdated, and probably gives inaccurate readings.
There is no electrical ground cable on the main!
The water shut-off valve is ancient and corroded.
The main water pipe is lead and must be replaced *(wiped joint noted by valve)*!
The water lines need to be properly secured to the wall. The wood board is not an acceptable support.

P 5. Electrical lines, conduits, and meters must be securely fastened to the side of the house. Tree branches need to be pruned away from the wires periodically. Three electrical lines at the service entrance head indicate 110/220 volts in this house. The "U" shape in the wires is called a drip loop. This is used to keep rainwater from entering the electrical conduit.

P 6. Caulking the joint on the top of the electrical meter and where the wires enter the house will prevent water penetration problems. Over time this exterior caulk will dry and crack and needs to be repaired.

The shingles on this house are made of asbestos/cement and are in excellent condition. EPA precautions must be taken when removing or tampering with any type of asbestos.

Home Inspection Photo Pages

P 7. In the lower level of many homes, you'll find wiring installed through the floor joists. This is acceptable as long as it meets the NEC requirements. Also, the drill holes must be in the center of the wood and less than 1/4 of the height of the beam. This will preserve the structural integrity of the wood beam.

You can see Romex and Bx cables are installed through these floor joists.

P 8. Using adapters when there is a lack of electrical outlets is a safety hazard. Too many appliances plugged into an outlet can create a fire or shock. The NEC recommends one outlet for every six feet of horizontal wall space. This will help prevent the use of usafe extension cord wiring and plug adapters.

To make matters even worse, this gas wall heater has a flexible supply line which is unsafe.

P 9. Here's a quiz: *How many problem conditions do you see in this photograph?*
1. There are remnants of asbestos that was unprofessionally removed from the old steam heating pipes.
2. The heating pipes and also between the floor joists should be insulated for energy efficiency.
3. The heating system flue stack has a downward pitch after the elbow which will slow the exhaust gases from exiting.
4. There is no pipe extending the water heater pressure relief valve to within eight inches above the floor.
5. On the lower, left of the tree trunk, the flexible pipe material is unsafe for the gas supply to the water heater.
6. This tree trunk could be put to better use somewhere else. In very old homes, you may find tree trunks being used to support the main girder beam. A solid, metal support post should be used instead.

P 10. A heating contractor took the easy route while installing this steam pipe. As a result, now there is a serious structural problem with the main girder beam. One-half of this beam was cut and removed which weakens the support. This pipe should have been routed around the girder. If that was not possible, then a hole, 1/4 of the height of the beam, could have been cut in the center of this girder.

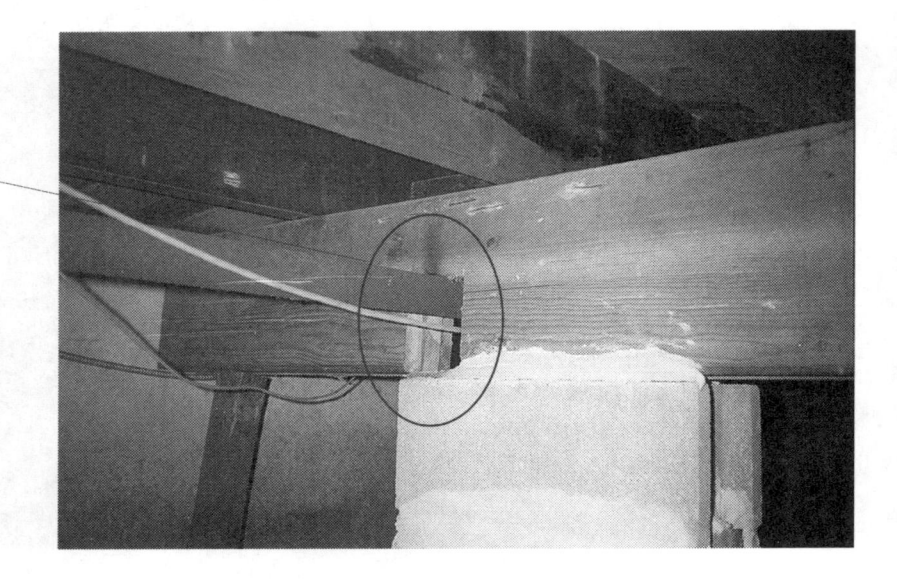

P 11. When inspecting the kitchen, spot check appliances by briefly turning them on. *Just remember to turn them off when you're done - except for the refrigerator!* The countertop and cabinets should be securely fastened. All outlets by the kitchen sink must have GFCI protection. Remodeled kitchens and bathrooms are like any other changes made to a house - building department permits and final approvals are neeed from town hall.

P 12. Bathroom tiles need to be evaluated for loose sections and open gaps. Loose areas can be detected by lightly banging on the tiles. The tiles in this bathroom are buckled and uneven. Prior water leaks behind the wall has caused this problem. To solve this problem, the tiles must be removed and the area behind must be repaired. Grout is used between ceramic tiles to prevent gaps that allow water penetration. Grout is a much harder material than caulk. Caulk is more flexible and used in areas where the joints will expand and contract more often.

P 13. This window is too close to the floor level and is a safety hazard. Not only can a child fall through, but if an adult tripped, there is no window sill to stop their fall. Child guards should be installed on this window. In many areas the height of the window sill and the use of child guards are regulated by the local building and fire codes. *Don't wait for accidents to happen - take precautions ahead of time.*
While you're inspecting the interior rooms, jump on the floors to make sure they're structurally sound.
Check underneath the corner of the wall-to-wall carpeting. The only way to know what type of flooring is underneath is to check it. Don't assume there are hardwood floors under carpets just because you see hardwood in other rooms in the house.
If there are any pets in the house, you should have the carpets removed or fumigated prior to moving-in. You don't want any fleas as house guests.

P 14. A vacuum seal is the airtight space between the panes of glass in thermal windows and doors. Broken vacuum seals are indicated by dirt and condensation stains in between the two panes of glass, such as in this sliding glass door. Over time the moisture and dust stains will increase. Since this area can't be wiped clean, the window will become white and hazy. Repairing broken vacuum seals is expensive.

Home Inspection Photo Pages				www.nemmar.com 97

P 15. Wood burning stoves can save a lot of money on heating fuel bills. These stoves can heat a large area of a home. However, safety precautions must be taken. Since these stoves radiate heat from the iron casing, they must not be touched while in use. A guardrail will help prevent accidental burns. Also, the flue stack for these stoves must be properly installed. A fireproof lining is needed and the flue should not be near any combustible materials, such as wood. Wood burning stoves, like forced hot air heating sytems, will dry out the air in the house. The metal pot on this stove is used to hold water. As the pot is heated, the water boils and turns to steam. This steam will add moisture back into the air so the occupants don't get sore throats or allergy problems from the dry air.

P 16. *Creosote* is a black soot found in chimneys. Creosote is caused by the smoke from burning wood. You may find excess creosote stains on the face of a chimney and mantle, such as this one. This indicates a backsmoking problem. Backsmoking is caused by a firebox area that is too narrow and/or a flue stack that does not extend high enough above the roof to prevent downdrafts.

P 17. *Here we have an accident waiting to happen!* Unfortunately, I have never seen any building codes that require railings around the attic access opening. Don't wait for someone to fall down this hole and break their neck *- install a guardrail NOW!* A handrail is also needed on the steps.

If plywood flooring is installed, the attic can provide storage space for lightweight items.

P 18. When inspecting the interior rooms make sure you look for signs of excessive structural settlement. Take a look at the floors where they meet the walls. You can see that this floor has separated from the baseboard molding due to abnormal settlement of the building over the years. It's a major expense to re-level and repair structural problems. Also, look at the window and door frames for signs of uneven and abnormal structural settlement.

P 19. *Here we are in asbestos heaven!* There are probably more asbestos fibers in this room than there are dust fibers. Almost always in older houses you'll find asbestos pipe insulation that is loose or has been removed unprofessionally. These conditions create very serious health hazards for the occupants of the house. Follow the EPA guidelines to resolve this.

P 20. Radon gas is considered by EPA to be the number two leading cause of lung cancer behind smoking. Radon is everywhere since it's created by a natural breakdown of rocks and soil. Stone foundation walls and dirt floors in the lower level increase radon gas levels. The large rock embedded in this basement will add radon into the air. A cement floor covering will help reduce this problem.

The insulation vapor barrier is installed upside down!

P 21. Asphalt/fiberglass shingles come in different weights. A heavier shingle has a 30 year life expectancy. Light weight shingles last about 20 years. These shingles are in good condition and there are no signs of old age or curling shingles.

With cable TV, antennas should be removed from roofs and chimneys. Antennas move in the wind and create water leaks. A cap and screen keep animals and water out of the chimney.

The small pipe in the roof is the vent stack for the plumbing drainage lines.

P 22. A new roof will be needed on this house soon. These asphalt shingles are old and at the end of their life expectancy. When the shingles cup and curl and get frayed edges, it's a clear sign of old age. Get estimates prior to buying this house since a roof can be a major expense. If there are two layers of shingles on the roof, a third layer should not be installed on top. Three layers are too heavy for the roof. Remove the prior two layers of shingles and check the condition of the plywood sheathing before adding the new layer of roof shingles.

P 23. All vines, ivy, shrubbery and trees must be pruned away from the house. This ivy clearly needs to be trimmed.

A minimum of at least eight inches above the soil is needed at the base of all siding. This clearance allows air and sunlight to help prevent rot and wood destroying insect problems.

Downspouts must be cleaned periodically. Clogged downspouts and gutters will create water problems around the foundation.

P 24. *Here's an example of what can happen if you don't read my books!* The wood siding on this garage has rotted at the base. This decay was caused by the wood touching the soil. An eight inch clearance between the soil and the base of the siding would have prevented this problem.

P 25. This welcome mat is resting on top of a safety hazard. Instead of "Welcome" this door mat should read: *"Stand here at your own risk!"*

A landing platforms is the standing area in front of a door. Landing platforms need to be large enough so the door can open safely. This storm door would knock someone down the steps if it was opened hastily.

The riser height is the vertical distance between each step. All risers should be evenly spaced about eight inches in height. This will help prevent tripping hazards from uneven stair heights.

P 26. Do you know what's missing in this picture? *(No, it's not a matching gargoyle that's missing).* A handrail needs to be installed on these stairs. Whenever there are more than two steps in height, a handrail is needed for safety.

All stairways should have a light to prevent tripping hazards at night.

P 27. There are several problems with this exterior deck support post.

1) The metal bracket is not sturdy enough to support the post where it meets the deck girder beam. Proper metal brackets need to be installed.

2) The base of this post is resting on a 2 x 6" wood board. This board will eventually rot and the deck will settle unevenly. Metal brackets are also needed at the base to properly secure the deck post to the concrete foundation.

P 28. Clearly these deck support posts are unsafe! This is a high deck and if it collapses, *then someone is going to get hurt or killed!!* The posts have been installed improperly. The base of the wood is not resting evenly on the concrete foundation. The second post in the photo is off center on the concrete foundation. Metal brackets are needed to securely fasten deck supports to the concrete base.

P 29. Are you tempted to just dive into this pool? Well don't because it's only a photo and you'll hit your head on the table where you're reading this book! *(Unless my book already put you to sleep).*

Check for any bulging or cracked sections of in-ground pools. The area around the pool, called the apron, should be smooth without any tripping hazards. Fences are required around all pools to prevent unattended children from entering the water. Flotation devices should be within reach in case of an emergency.

P 30. Pools need to be winterized properly to prevent the walls from cracking due to freezing water during cold months.

P 31. Here we have the exposed termite damage in a lower level floor joist. This is a common area for damage since the wood is close to the soil. The termite channels can be seen when the wood is probed and opens up. The mud and termite tunnels are mostly hidden from the light under a thin, outer layer of the wood. This is because termites need a dark, warm and moist environment at all times.

P 32. On the exterior check the wood trim near the roofline. This area is prone to carpenter bee damage. The holes in this fascia board are an example of a carpenter bee nest. Carpenter bees and carpenter ants do not eat the wood for food, they merely excavate it to make a nest. Keeping a solid coat of stain or paint can reduce the chance of wood destroying insect damage.

More Nemmar Products

Email info@nemmar.com for prices

Energy Saving Home Improvements From A to Z ™

Don't let your dream house be a money pit in disguise! Our **5-star rated** book that teaches you how to **save** thousands of dollars **and** help the environment by making minor improvements to your home. You'll learn how to **lower your utility bills by 50%,** live more comfortably, and help the environment. Includes many photographs with detailed descriptions.

Home Inspection Business From A to Z ™

The REAL FACTS the other books don't tell you! Our **number one** selling home inspection book. This is **definitely** the best home inspection book on the market and has been called the "Bible" of the inspection industry. *Every* aspect of home inspections is covered with precise steps to follow. Includes many photographs with detailed descriptions.

Real Estate Appraisal From A to Z ™

The REAL FACTS the other books don't tell you! Our **number one** selling appraisal book. This is **definitely** the best real estate appraisal book on the market. *Every* aspect of real estate appraising is covered with precise steps to follow. Includes sample professional appraisal reports and many photographs with detailed descriptions.

DVD's - Home Inspection From A to Z ™

Our **5 star rated** DVD's have two hours of video plus you get the 80 page *HIB **DVD** Companion Guidebook!*
OPERATING SYSTEMS DVD topics including: heating systems, air-conditioning, water heaters, plumbing, well water system, septic system, electrical system, gas service, and auxiliary systems. Health Concerns topics including: asbestos insulation, radon gas, and water testing.
INTERIOR and EXTERIOR DVD topics including: roof, chimneys, siding, eaves, gutters, drainage and grading, windows, walkways, entrances and porches, driveways, walls and fences, patios and terraces, decks, swimming pools, exterior structures, wood destroying insects, garage, kitchen, bathrooms, floors and stairs, walls and ceilings, windows and doors, fireplaces, attics, ventilation, insulation, basement/lower level, and water penetration.

Home Buyer's Survival Kit ™

Don't buy, sell, or renovate your home without this! Includes: Four of our **top selling** books – *Real Estate Home Inspection Checklist From A to Z, Energy Saving Home Improvements From A to Z, Home Inspection Business From A to Z,* and *Real Estate Appraisal From A to Z.* Plus, you get **both** of our *Home Inspection From A to Z* – **DVD's.** As an added bonus you also get the 80 page *HIB **DVD** Companion Guidebook.*

Narrative Report Generator and *On-Site Checklist*

The report generator and checklist the others don't have! CD-Rom with the *best* Narrative Report Generator and On-Site Checklist on the market! These will enable you to *easily* do 30 page narrative, professional home inspection reports to send to your clients. These will assist you at the inspection site to be sure that you properly evaluate the subject property. Designed to walk you through the entire inspection process with very detailed instructions on how to properly evaluate the condition and status of **all** aspects of a home in a fool-proof, step-by-step system and create professional, narrative reports.

Just some of our books, CD's, DVD's and much more!
Email **info@nemmar.com** for prices.
Visit us at **www.nemmar.com**

Subject Property #2

◊ The address of the subject property is _____.

◊ The inspection was commissioned by _____.

◊ The date of the inspection was _____.

◊ The time of the inspection was from _____ to _____ AM PM.

◊ The outdoor air temperature was approximately _____ degrees.

◊ The weather on the day of the inspection was (raining/sunny)_____.

◊ *You must use and read the entire inspection report to get the maximum benefit!* Do not just take it for granted that everything is in good condition at the subject property. The written inspection report has valuable and important information that you need to know in order to properly evaluate the subject property.

◊ A home inspection is a visual, limited time, non-destructive, and non-dismantling inspection. There is no dismantling or using tools to take things apart. There are areas that are inaccessible or not visible, such as behind finished wall, floor and ceiling coverings, etc.

◊ A home inspection checks to see if all visible and accessible areas and operating systems, such as, heating, air-conditioning, electrical, plumbing, roof, etc. are operating properly at the time of the inspection. The inspection tries to determine what the current condition and life expectancy is of the different aspects of the subject property. The inspection is limited in evaluating the life expectancies of an item without knowing what the past maintenance history was for the item being evaluated.

◊ During the inspection you cannot turn on or test any devices, appliances, operating systems, electrical switches, etc., that do not operate by the normal controls used to operate it that were designed by the manufacturer. For example, you can only use thermostats to test the heating or air-conditioning systems, etc. You cannot test anything that has *"do-it-yourself"* wiring and installations. Any do-it-yourself type of setups can be dangerous to operate and are not part of the home inspection process.

◊ It is *highly* recommended that any problem conditions noted at the on-site inspection or in this written report be evaluated by a reputable, licensed contractor **PRIOR** to signing any contracts or closing on the subject property. The inspection performed is not a building code inspection. It is also recommended that you check *all* records at town hall pertaining to the subject property.

Questions To Ask The Home Seller #2

When you ask these preinspection questions, make sure that you ask the owner and Realtor about information from any prior owners of the house. Meaning that if the seller tells you,
"No, we have never made any changes to the foundation or septic system,"
then ask them if they know of any <u>prior</u> owners having made any changes, repairs, etc.

◊ *Age, Zoning, and Permits:*

◊ What is the age of the house/condo? _____

◊ How long you lived in this house/condo? _____

◊ Are there any outstanding building, zoning or other violations or any missing permits and/or approvals?
_____

◊ *Interior Inspection:*

◊ Are there any damaged areas to the floors, walls, and/or ceilings that you know about? _____

◊ If yes, then where are these damaged areas and are they hidden by carpets, furniture, sheetrock, etc.?
_____

◊ Has any insulation been added or removed in the floors, walls, and/or ceilings? _____

◊ If yes, then explain the details of the insulation added or removed: _____
_____

◊ Has any UFFI foam or asbestos insulation been removed from the house and if yes, then do you have the licensed EPA contractor certification for that work? _____

◊ Does the fireplace draft properly or are there back-smoking problems and how often do you use the fireplace?
_____

◊ *Exterior Inspection:*

◊ Has there been any exterior siding added to the house after the original construction? _____
_____

◊ Do you know what type of insulation and materials are behind the exterior siding between the walls?
_____

◊ What is the age of the roof? _____

◊ Do you know how many layers of shingles there are on the roof? _____

◊ Have there been any past or present water leaks or problems with the roof? _____
_____

◊ Have any decks, additions or updating been done? If yes, what are the details of that work and are all valid permits and Certificate of Occupancies, *(C of O)*, filed at town hall? _____
_____

◊ Have any structural renovations been done? If yes, is there a valid permit and C of O for the work done?
_____

◊ Have there been any structural problems in the house? If yes, what are the details? _____
_____

◊ **_Operating Systems:_**

◊ Can I test all operating systems in the house or are there any that are being repaired or aren't functioning properly? *(Operating Systems refers to items such as the heating, air-conditioning, plumbing, electrical, wells, septics, etc.)* _____
_____

◊ Do you know the age of the furnace/boiler? _____

◊ How often and what company services/maintains the heating system? _____

◊ Are all the rooms in the house heated? _____

◊ Are there any oil tanks, used or unused, on the property? _____

◊ If yes, where are those oil tanks located and what is the age of those oil tanks? _____
_____

◊ Have there been any problems with the heating system in the house? If yes, what are the details? _____
_____

◊ Do you know the age of the air-conditioning compressor? _____

◊ Did the air-conditioning system operate properly last season? *(if it's too cold to test it now)* _____

◊   How often and what company services/maintains the A/C system? _____

◊   Have there been any problems with the air-conditioning system in the house? If yes, what are the details? _____

◊   Is the house/condo connected to municipal water & sewer systems? *(This is very important to get from them since there is no way to determine this at the site without checking the town hall records).* _____
_____

◊   Has there been any past or present problems with the water pressure and drainage in the house plumbing system? _____
_____

◊   Has there been any past or present problems with electrical overloads, outlets, switches, etc.? _____
_____

◊   *Termite and Water Problems:*

◊   Has the house ever had termites or Wood Destroying Insect damage? _____

◊   Has the house ever been treated to remove or prevent termites or Wood Destroying Insects? _____

◊   If yes, when and what are the details of that WDI treatment? Are there any guarantees or documentation for the WDI treatments? _____
_____

◊   Have there been any water penetration problems in the house? If yes, what are the details? _____
_____

◊   Are there any sump pumps in the lower level area to remove water? _____

◊   *Septic System:*

◊   Is there a survey or plot plan showing the septic system? _____

◊   Have there been any renovations or additions to the house needing septic system approvals, such as bathrooms/bedrooms added? _____

◊   Have there been any past or present problems with the septic system? _____
_____

◊   Do you know the location of the septic tank and leaching fields? _____

◊   Do you know what the size of the septic tank is? _____

◊   Do you know what construction materials the septic tank is made of? _____

◊ Is the septic tank original or was it upgraded or replaced over the years? _____

◊ When was the septic tank last pumped out and internally inspected? _____

◊ How often has the septic tank been pumped out and internally inspected over the years prior to the last cleaning? _____

◊ What is the name of the septic service company that maintains the system? _____

_____

◊ **_Well Water System:_**

◊ Are there any surveys or plot plans showing the well water system? _____

◊ What is the depth of the well? _____

◊ Is the well water pressure and volume adequate for normal use? _____

◊ Has there been any past or present problems with the well water pressure and volume? _____

_____

◊ When was the well pump last serviced or replaced? _____

◊ When was the well water storage tank last serviced? _____

◊ What is the age of the well water storage tank? _____

◊ What is the name of the well water service company that maintains the system? _____

_____

◊ **_Swimming Pool:_**

◊ What is the age of the pool, filter, heater and liner? _____

◊ Do you have a Certificate of Occupancy and all valid permits for the pool? _____

◊ Have there been any leaks in the pool walls or other problems with the pool or pool equipment? _____

_____

◊ Has the pool been properly winterized? *(if applicable)* _____

◊ What is the name of the service company that maintains the swimming pool? _____

_____

_____

The Operating Systems Inspection #2

Heating System

◊ The brand name of the heating system installed is _____.

◊ The fuel for the heating system is (gas/oil/electricity)_____.

◊ The overall BTU capacity of the heating system is _____.

◊ The age of the heating system is approximately (#)_____ years. The life expectancy is generally (#)_____ years for this type of heating system.

◊ The last date of service/repairs for the heating system was _____.

◊ The ceiling and walls around the heating system should have a covering of sheet-metal or 5/8 inch fireproof sheetrock to help prevent the spread of fires in this area. The fireproof covering is (y/n)_____ installed.

◊ The flue pipe sections are in (G/A/F/P)_____ condition. The joints at the connecting sections do (y/n)_____ have the required screws to keep them in place. The flue pipe does (y/n)_____ have the required upward pitch. These are all required items that must be installed. The flue pipe is used to safely discharge the carbon monoxide and other products of combustion that are caused by gas and oil fired burners. These gases **must be safely discharged** from the house. They are ***LETHAL GASES!!***

◊ The flue pipe is (y/n)_____ within 4 inches of any combustible material, such as wood. A minimum clearance of 4 inches is required to help prevent fires.

◊ The heating system is operated by (#)_____ zone(s). A zone is an area of the house with a separate thermostat which can have a different setting. Having more than one heating zone is more energy efficient.

◊ The heating system was tested by turning up all zone thermostats to engage the heating system for about (#)_____ minutes. All radiators/registers did (y/n)_____ operate properly during the testing by getting warm.

◊ The emergency shutoff switch did (y/n)_____ operate properly when tested. This is used to shut the system off during repairs and for emergencies by overriding the thermostat control.

_____
_____
_____
_____

◊ ***Oil Fired Heating Systems:***

◊ Oil fired systems *must* be tuned up every year by a reputable heating service contractor for efficient operation. Most oil delivery companies will provide a service contract with the Owner to service and tune up the oil burner and provide emergency repairs and maintain the oil tanks. The burner flame must be adjusted every year, the oil filters changed, the flue draft regulator adjusted, and the flue pipe cleaned.

◊ When the oil burner engaged, there were (y/n)_____ signs of back-smoking. Any back-smoking will indicate a problem condition that needs to be repaired.

◊ The oil burner flame was (y/n)_____ able to be viewed for proper color and height.

◊ The firebox is in (G/A/F/P)_____ condition. This area needs to be periodically monitored due to deterioration from the high temperature of the burner flame.

◊ The oil feed lines are in (G/A/F/P)_____ condition. The feed lines should be made of copper and should be covered to prevent them from being damaged or becoming a tripping hazard.

◊ There is (y/n)_____ a required firematic shutoff valve within 6 feet of the burner for safety to shut off the fuel. The purpose of this is to be able to shut off the flow of oil if it is necessary. This is different from the emergency shutoff switch because it does not turn off the burner, it only shuts off the oil supply.

◊ The oil filter is in (G/A/F/P)_____ condition.

◊ The oil tank is located _____. The oil tank was (y/n)_____ able to be viewed for rusting or corrosion problems. It is *highly* recommended that the oil tank be tested by a reputable oil contractor to determine if there are any present leaks or potential leaks due to corroded sections of the tank.

◊ Oil tanks generally last about 25 to 30 years and longer if maintained properly. Determine from the Owner or the oil contractor what the exact age of the oil tank(s) is. Determine if any Certificate of Occupancy, permits or surveys are needed in the local municipality with on-site oil tanks.

◊ The draft regulator on the flue pipe is in (G/A/F/P)_____ condition. This needs to be adjusted every year with the tune-up of the heating system. The purpose of it is to allow some cool air from the boiler room area to help with the removal of carbon monoxide up the chimney, without drawing too much heat from the boiler or furnace.

_____

_____

_____

_____

_____

_____

The Operating Systems Inspection #2   www.nemmar.com   115

◊ **_Gas Fired Heating Systems:_**

◊ There is (y/n)_____ a required gas shutoff valve within 6 feet of the burner for safety to shut off the fuel. The purpose of this is to be able to shut off the flow of gas if it is necessary. This is different from the emergency shutoff switch because it does not turn off the burner, it only shuts off the gas supply.

◊ The gas feed lines do (y/n)_____ appear to be made of approved piping. The gas lines <u>MUST</u> have approved black iron gas piping for the feed lines and not copper or other materials that are not approved to carry gas fuel.

◊ The burner flames were (y/n)_____ able to be checked for proper color and height. They should be checked periodically and should be as blue as possible with very little yellow or orange color. Too much yellow or orange color means that the fuel and air mixture needs to be adjusted.

◊ The inspector was (y/n)_____ able to view the heat exchanger. There were (y/n)_____ signs of excessive rust or cracks in the visible and accessible areas. Sealed systems cannot be fully inspected due to lack of access.

◊ The draft diverter hood at the base of the flue pipe is in (G/A/F/P)_____ condition. It is used to keep downdrafts in the chimney from blowing out the pilot light and to help keep the heat inside the boiler or furnace while the carbon monoxide is removed.

_____
_____
_____
_____
_____
_____

◊ **_Electric Heating Systems:_**

◊ The Owner or Realtor stated that there is (y/n)_____ problem with blown fuses or tripped circuit breakers due to the additional electrical usage for the electrical heating system.

◊ All of the electric baseboard radiators were (y/n)_____ warm during the testing. The radiators are in (G/A/F/P)_____ condition.

_____
_____
_____

◊ **_Forced Warm Air Systems:_**

◊ There is (y/n)_____ at least one supply vent in each room providing heat.

◊ The air filters were (y/n)_____ clean when checked. They need to be replaced every few months during the heating season. This is similar to changing your car air and oil filters. If you do not do it often enough, you will create excess wear and tear on the furnace due to the lack of maintenance.

◊ The air ducts are (y/n)_____ insulated. All ducts should be insulated, internally or externally, for maximum energy efficiency.

◊ There is (y/n)_____ a humidifier on the heating ducts. Forced hot air heating systems will dry out the air in a house and can lead to the occupants getting sore throats. A humidifier will help prevent this. Humidifiers cannot be fully tested during a home inspection due to their operation. The life expectancy of a humidifier is generally about 5 to 7 years.

◊ The furnace plenum is (y/n)_____ separated from the heat exchanger and fan area by a canvas or flexible type of material. This will help prevent any vibrations caused by the blower fan from being transmitted through the ducts and into the livable rooms.

◊ The inspector was (y/n)_____ able to view the heat exchanger and the fan. There were (y/n)_____ signs of excessive rust or cracks in the visible and accessible areas. Sealed systems cannot be fully inspected due to lack of access. If the heat exchanger is cracked it will leak carbon monoxide into the supply ducts of the house. If there is a leak of any kind it must be checked out **_IMMEDIATELY_** before using the heating system again. The carbon monoxide that is released is a **LETHAL GAS!!**

◊ The fan was (y/n)_____ making unusual noises while it was operating. Abnormal or unusual noises while operating will indicate repairs may be needed.

_____
_____
_____
_____
_____
_____
_____
_____
_____
_____

◊ **_Heat Pump Heating Systems:_**

◊   The Owner or Realtor stated that the heat pump unit does (y/n)_____ heat the house adequately in the cold months.

◊   Heat pumps cannot be tested in both the heating and the air-conditioning modes during an inspection. Doing so can damage the compressor unit. When it is working properly in one mode, then it is an indication that the most important and costly parts are operating properly.

◊   The age of the compressor unit is approximately (#)_____ years. The life expectancy of a compressor is about 10 to 12 years, depending upon the amount of usage and maintenance given to the system over the years.

◊   The compressor is (y/n)_____ resting on a sturdy, level foundation, like concrete. Uneven installations can cause premature failure of the compressor because it will be leaning to one side while operating.

◊   The compressor was (y/n)_____ making unusual noises while it was operating. Abnormal or unusual noises while operating will indicate repairs may be needed.

◊   There is (y/n)_____ a required exterior service disconnect switch next to the compressor for emergency and repairs shut off.

◊   There was (y/n)_____ adequate ventilation around the compressor unit for the air intake and blower fan to operate properly. All trees and bushes should be pruned away at all times and there should be no obstructions overhead.

◊   The compressor coils were (y/n)_____ clean when checked. The coils need to be cleaned periodically for proper maintenance and operation of the system.

_____
_____
_____
_____
_____
_____
_____
_____
_____
_____
_____

◊ **_Forced Hot Water Heating Systems:_**

◊ There is (y/n)_____ at least one radiator in each room for heating purposes.

◊ The circulator pump(s) is in (G/A/F/P)_____ condition and was (y/n)_____ operating properly at the time of the inspection. Circulator pumps need just a drop of oil in the oil ports approximately once a year. Have the heating service contractor check this with the annual tune ups.

◊ The visible heating pipes and pipe joints are in (G/A/F/P)_____ condition. There are (y/n)_____ signs of excessive rust or leaking conditions. Any rust or leaking conditions will require repairs by a licensed contractor.

◊ The heat exchanger was (y/n)_____ able to be viewed for signs of excessive rust or leaking conditions.

◊ The water pressure reducing valve is in (G/A/F/P)_____ condition. This reduces the water pressure that is coming from the house plumbing lines, which is usually about 30 to 60 psi *(pounds per square inch),* down to about 12 to 15 psi before entering the boiler.

◊ There is (y/n)_____ a required backflow preventer next to the water pressure reducing valve. This prevents water that has entered the boiler from re-circulating back into the house plumbing lines and mixing with the faucet and shower water supply which is a health hazard.

◊ The expansion tank is in (G/A/F/P)_____ condition. When you heat water it expands. The heated water needs a cushion to expand or else it will burst some of the pipe joints or discharge the pressure relief valve to relieve the pressure. The expansion tank has an air pocket or a rubber bag in it that cushions the water as it expands so the pressure in the system does not get too high. It needs to be checked periodically for water-logging problems. It should have a bleeder valve to put air in it if it becomes waterlogged. It should have a drainage valve on the bottom to drain it at least once a year for any rust and sediment that builds up in the tank.

◊ The pressure gauge does (y/n)_____ appear to be operating properly. The proper operating pressure for a hot water heating system is between 12 to 22 psi.

◊ The pressure relief valve is (y/n)_____ located directly on the boiler for safety. This valve is in (G/A/F/P)_____ condition. This is a safety device that helps prevent the heating system from becoming dangerously high in pressure. If the pressure reaches 30 psi then the valve will discharge to relieve the system pressure so the boiler will not explode.

◊ The pressure relief valve is (y/n)_____ piped properly for safety. It *must* be piped to within 8 inches of the floor to prevent scalding anyone when discharging. When water or steam discharges from this valve, it indicates a problem condition and the system *must* be checked out immediately by a licensed heating contractor.

◊ There is (y/n)_____ a drain valve on the lower part of the boiler. This is used to drain some water into a bucket each month to remove some of the rust and sediment that normally builds up in the system.

_____

_____

_____

_____

◊ **_Steam Heating Systems:_**

◊ There is (y/n)_____ at least one radiator in each room for heating purposes.

◊ The visible heating pipes and pipe joints are in (G/A/F/P)_____ condition. There are (y/n)_____ signs of excessive rust or leaking conditions. Any rust or leaking conditions will require repairs by a licensed contractor.

◊ The heat exchanger was (y/n)_____ able to be viewed for signs of excessive rust or leaking conditions.

◊ The pressure gauge does (y/n)_____ appear to be operating properly. The proper operating pressure for a steam heating system is between 2 to 5 pounds per square inch, *(psi)*.

◊ The upper limit switch is (y/n)_____ located directly on the boiler for safety. This switch is in (G/A/F/P)_____ condition. If the pressure gets too high, this switch will turn off the burner.

◊ The pressure relief valve is (y/n)_____ located directly on the boiler for safety. This valve is in (G/A/F/P)_____ condition. This is a safety device that helps prevent the heating system from becoming dangerously high in pressure. If the pressure reaches 15 psi then the valve will discharge to relieve the system pressure so the boiler will not explode.

◊ The pressure relief valve is (y/n)_____ piped properly for safety. It must be piped to within 8 inches of the floor to prevent scalding when discharging. When water or steam discharges from this valve, it indicates a problem condition and the system *must* be checked out immediately by a licensed heating contractor.

◊ There is (y/n)_____ a drain valve on the lower part of the boiler. This is used to drain some water into a bucket each month to remove some of the rust and sediment that normally builds up in the system.

◊ The water level in the boiler sight glass was (y/n)_____ at the proper level at the time of the inspection. The sight glass water level should be 1/2 to 3/4 of the way full. It should not be completely empty nor completely full which will cause problems in the system. The sight glass level allows you to see that there is air and water in the boiler. This is because you are making steam and you have got to have room for the heated water to boil and create the steam.

_____
_____
_____
_____
_____
_____
_____
_____

Air-Conditioning System

◊ There is (y/n)_____ a central air-conditioning system for the subject property.

◊ The Owner or Realtor stated that the unit does (y/n)_____ cool the house adequately in the warmer months. An air-conditioning system is a closed system, and, theoretically, there should never be a need for additional Freon. However, in practice, the various fittings on the connecting pipes can loosen or develop hairline cracks that can allow some of the Freon gas to escape. If the air-conditioning system cannot hold a Freon charge for at least one season, then the leaks in the pipes or fittings should be located and sealed.

◊ The outdoor air temperature was (y/n)_____ warm enough to test the air-conditioner at the time of the inspection. Central or window air-conditioning units **cannot** be tested when the outdoor air temperature is 65 degree Fahrenheit or lower. The interior pressure that is required to properly operate an air-conditioning system is too low when the outdoor air temperature is 65 degrees or lower.

◊ All window and wall air-conditioning units were (y/n)_____ able to be spot checked for proper operation. They did (y/n)_____ operate properly when tested. Determine from the Owner or Realtor if the portable air-conditioning units are being sold with the house.

◊ The air-conditioning system was tested by turning up all of the zone thermostats to engage the system for about (#)_____ minutes. All vents/registers did (y/n)_____ operate properly during the testing by discharging cool air.

◊ The inspector was (y/n)_____ able to spot check a supply vent with a thermometer to determine if the discharging air was cool enough. The reading should be about 55 to 58 degrees Fahrenheit. The temperature reading noted was (#)_____ degrees at the time of the inspection.

◊ There is (y/n)_____ at least one supply vent in each room providing cool air.

◊ The inspector was (y/n)_____ able to view the air filters. The air filters were (y/n)_____ clean when checked. They need to be replaced every few months during the air-conditioning season. This is similar to changing your car air and oil filters. If you do not do it often enough, you will create excess wear and tear due to the lack of maintenance.

◊ The air ducts are (y/n)_____ insulated. All ducts should be insulated, internally or externally, for maximum energy efficiency.

◊ The air handler plenum is (y/n)_____ separated from the fan area by a canvas or flexible type of material. This will help prevent any vibrations caused by the blower fan from being transmitted through the ducts and into the livable rooms.

◊ The inspector was (y/n)_____ able to view the evaporator coil and the fan. There were (y/n)_____ signs of excessive rust or cracks in the visible and accessible areas that need to be evaluated by a licensed contractor. Sealed systems cannot be fully inspected due to lack of access.

◊ The inspector was (y/n)_____ able to view the condensation drain pan under the evaporator coil. The condensation drain pan is in (G/A/F/P)_____ condition. The drain pan drainage line does (y/n)_____ lead to a condensate pump that removes the condensation to a suitable location. The life expectancy of these pumps is about 5 to 7 years.

◊ The fan was (y/n)_____ making unusual noises while it was operating. Abnormal or unusual noises while operating will indicate repairs may be needed.

◊   The age of the compressor unit is approximately (#)_____ years. The life expectancy of an air-conditioning compressor is about 10 to 12 years, depending upon the amount of usage and maintenance given to the system over the years.

◊   The compressor is (y/n)_____ resting on a sturdy and level foundation, such as a concrete base. Uneven installations can cause premature failure of the compressor because it will be leaning to one side while operating.

◊   The compressor was (y/n)_____ making unusual noises while it was operating. Abnormal or unusual noises while operating will indicate repairs may be needed.

◊   There is (y/n)_____ a required exterior service disconnect switch next to the compressor for emergency and repairs shut off. There was (y/n)_____ adequate ventilation around the compressor unit for the air intake and blower fan to operate properly. All trees and bushes should be pruned away at all times and there should be no obstructions overhead. The coils were (y/n)_____ clean when checked. The coils need to be cleaned periodically for proper maintenance of the system.

◊   The high and low pressure lines were checked where visible and accessible. The lines are (y/n)_____ made of copper. The low pressure line, which is the larger pipe that is about 3/4 inch in diameter, is (y/n)_____ insulated. This line must be insulated for energy efficiency since it has cold Freon in it. The high pressure line is the thin diameter line and it does not need to be insulated.

_____
_____
_____
_____
_____
_____
_____
_____
_____
_____
_____
_____
_____
_____
_____

Domestic Water Heater

◊ *<u>Separate Domestic Water Heaters:</u>*

◊ The domestic water heater is fueled by (gas/oil/electricity)_____.

◊ The capacity of the water heater is (#)_____ gallons. The standard size water heater for a single family house is 40 gallons. Oversized water heaters in the house are not as energy efficient because a lot of water will just sit in the tank and will not be used after being heated.

◊ There were (y/n)_____ signs of excessive rust or water leaking conditions on the unit. Any rust or leaking conditions will require repairs by a licensed contractor.

◊ The temperature/pressure relief valve is (y/n)_____ located directly on the water heater for safety. This valve is in (G/A/F/P)_____ condition. This is a safety device that helps prevent the water heater from becoming dangerously high in temperature or pressure. If the temperature reaches 210 degrees Fahrenheit or the pressure reaches 150 psi, then the valve will discharge to relieve the system so the water heater will not explode.

◊ The pressure relief valve is (y/n)_____ piped properly for safety. It *must* be piped to within 8 inches of the floor to prevent scalding anyone when discharging. When water or steam discharges from this valve, it indicates a problem condition and the system *must* be checked out immediately by a licensed contractor.

◊ There is (y/n)_____ a drain valve on the lower part of the water heater. This is used to drain some water into a bucket each month to remove some of the rust and sediment that normally builds up in the system.

◊ The hot and cold water lines do (y/n)_____ appear to have a reversed installation *(hot pipe in cold pipe slot)*. A reversed installation with the hot and cold lines will cut down the energy efficiency of the water heater.

◊ The estimated age of the water heater is (#)_____ years. The life expectancy of a water heater is 10 to 12 years depending upon the maintenance given to it over the years.

◊ The water heater temperature setting was (#) or (hot/warm/cold)_____ at the time of the inspection. All water heaters should be kept on the "warm" setting for maximum efficiency and life expectancy. The warm setting on the water heater thermostat is usually about 125-130 degrees Fahrenheit which is the factory recommended temperature setting for most water heaters. A high temperature setting will cause the unit to constantly heat the water higher than is necessary, which can cause excessive wear and tear and premature failure.

◊ During the interior inspection the faucets were spot checked and adequate hot water was (y/n)_____ available.

_____
_____
_____
_____
_____
_____

◊ **_Immersion Coil Water Heaters:_**

◊ The domestic hot water is supplied by an immersion coil system inside the boiler. An immersion coil system has water pipes that carry cold water inside of a coil located in the side of the boiler. The coils are *"immersed"* in the hot boiler water, hence you get the name *"immersion coils."* The cold water in the pipes **should not** mix with the boiler water because then the dirty boiler water would be carried back to the faucets and showers, which would be a health problem.

◊ The advantage of an immersion coil system is that it provides inexpensive hot water in the winter time because the boiler is already operating to heat the house. The disadvantage of an immersion coil system is that it is not as energy efficient as having a separate water heater unit in the warmer months. It also adds an unwanted heat load on the house in the warmer months. Also, the coils clog over time and need to be cleaned periodically.

◊ During the interior inspection the faucets were spot checked and adequate hot water was (y/n)_____ available.

◊ **_Oil Fired Water Heaters:_**

◊ Oil fired units have a very fast recovery rate, which is the rate at which they can re-heat the water. There is (y/n)_____ a required water temperature setting switch. The purpose of this is to operate like a thermostat to regulate the burner to turn on and off to keep the water at a pre-set temperature. The factory recommended setting is usually at 125-130 degrees Fahrenheit. If it is set too high, then someone can get scalded with very hot water.

_____

_____

_____

_____

_____

◊ **_Gas Fired Water Heaters:_**

◊ There is (y/n)_____ a required water temperature setting switch. The purpose of this is to operate like a thermostat to regulate the burner to turn on and off to keep the water at a pre-set temperature.

_____

_____

_____

_____

_____

◊ **_Electrically Operated Water Heaters:_**

◊ Electrically operated units have coils that are directly immersed in the water inside the tank. They usually have two switches inside the small cover plates on the side of the tank to regulate the water temperature setting.

_____

_____

_____

_____

_____

The Operating Systems Inspection #2   www.nemmar.com

Plumbing System

◊ The visible plumbing lines and joints were in (G/A/F/P)_____ condition. There were (y/n)_____ signs of excessive rust or leaks that need to be further evaluated by a licensed plumbing contractor. The types of plumbing line materials used in housing construction include: Copper, Brass, Galvanized Iron, Lead, PVC, and Cast Iron. The visible **supply** plumbing lines are made of _____. The visible **drainage** plumbing lines are made of _____.

◊ The visible portion of the water main entry line is made of _____. The main water line is (y/n)_____ securely fastened to the wall to help prevent damage to the plumbing joints.

◊ The water main shutoff valve is in (G/A/F/P)_____ condition. This valve does (y/n)_____ appear to be in proper working order. This valve is used to turn off **all** of the water entering the house in case of an emergency or if any repairs are being performed. The main shutoff valve cannot be tested during a home inspection due to the possibility of it becoming rusty over time and if tested it can *"freeze"* in the on or off position, or possibly leak.

◊ There is (y/n)_____ a water meter reading device installed on the main water line. This is used by the utility company to calculate the water usage bill for the homeowner.

◊ There is (y/n)_____ an electrical grounding wire on the water main line for the electrical system. **This is a very important safety item and must not be disconnected or rusty!!** The grounding wire and clamps **must** be checked periodically for any rust or corrosion. The purpose of this wire is to ground the electrical system for safety. The grounding wire does (y/n)_____ span the water meter and shutoff valve. The grounding wire should be clamped on both sides of the water meter and the shutoff valve with a *"jumper cable"* for safety.

◊ There is (y/n)_____ a pressure reducing valve installed by the water meter. This valve is generally found in areas where the municipal water system has very good pressure from the street water lines.

◊ During the interior inspection, the inspector spot checked the water pressure and drainage by briefly running the faucets. The water pressure and drainage was (y/n)_____ in proper working order. The client was (y/n)_____ present to view this testing.

_____
_____
_____
_____
_____
_____
_____

Well Water System

◊ There is (y/n)_____ a well water system for the subject property.

◊ The inspector is very limited in evaluating a well system because most of the components are underground and/or not visible, such as the well pump, water lines and sometimes even the water storage tank is located in an underground pit. Also, the inspector has no accurate knowledge of what the past maintenance history has been for the well and the well equipment.

◊ It is *highly* recommended that you obtain all building department permits, surveys, plot plans and approvals for the well system. Most wells are deep and the repair costs and fees to drill deep wells are **much higher** than shallow wells because the price is usually based on the length of water piping used and the drilling depth.

◊ It is *highly* recommended that you have a water sample taken and analyzed at a reputable laboratory whenever the house has a well water system or possible lead plumbing lines. Laboratory water analysis should be done to test for bacteria, mineral, metal and radon content in the water supply.

◊ The well pump is located inside the well and is not visible. The age of the well pump is reported to be (#)_____ years. The life expectancy of a well pump is about 7 to 10 years, but can be longer if it is not overworked or neglected. The life expectancy also depends upon the type and quality of the pump installed and the acidity of the well water.

◊ The well pressure gauge does (y/n)_____ appear to be operating properly. Pressure gauges often get rusty and need to be replaced every few years.

◊ The water storage tank is in (G/A/F/P)_____ condition. There were (y/n)_____ signs of excessive rust on the tank that need to be evaluated further by a licensed well contractor. The tank is (y/n)_____ insulated to help prevent rust from condensation. The age of the storage tank is approximately (#)_____ years. The life expectancy of a water storage tank is about 15 to 20 years, and similar to a well pump, it depends upon the maintenance done, the type and quality of the tank and the acidity of the well water.

◊ There is (y/n)_____ a pressure relief valve for the well storage tank. This *must* be installed for safety in case the pressure in the system gets too high. It is usually set at 75 psi, *(pounds per square inch),* depending upon the type and capacity of the storage tank. The tank does (y/n)_____ have an air fill valve to adjust the air-to-water ratio inside of it during periodic maintenance of the system.

◊ The visible well water lines are in (G/A/F/P)_____ condition. There were (y/n)_____ signs of excessive rust on the lines. The lines are (y/n)_____ insulated to help prevent rust from condensation.

◊ The well water system was tested by turning on (#)_____ faucets to engage the system for about (#)_____ minutes. All faucets did (y/n)_____ operate properly during the testing by providing an adequate flow of water. The water flow at the time of the inspection was approximately (#)_____ gallons per minute. The minimum acceptable flow for a well system is 5 gallons per minute, *(GPM)*. Some local area codes may require a higher GPM rating. The client was (y/n)_____ present to view the testing.

◊ The "idle" or "static" pressure reading was (#)_____ psi before the well testing began. During the well water test the pressure gauge had a high reading of (#)_____ psi and a low reading of (#)_____ psi. The well pressure should remain within a 20 psi differential during and after the test. This simply means that the high and low pressure gauge reading of the system should not be more than a 20 psi difference during use.

◊ There is (y/n)_____ a properly operating emergency shutoff switch for the well pump.

◊   There is (y/n)_____ a water filter system installed. There is (y/n)_____ a water softener installed. The inspector cannot evaluate water filtration or water softener systems during a home inspection because of the laboratory water analysis that would be needed. The water filters and brine need to be replaced according to the owners manuals and whenever they appear dirty.

◊   If a water softener is installed and the house is serviced by a septic system, then the brine water from the softener should not discharge into the septic system. The brine water will alter the natural bacterial action of the septic decomposition inside the septic tank and can cause premature failure of the septic system.

Septic System

◊ There is (y/n)_____ a septic system for the subject property.

◊ The inspector is very limited in evaluating a septic system because most of the components are underground and/or not visible, such as the drainage lines, the holding tank and the leaching fields or seepage pits. Also, the inspector has no accurate knowledge of what the past maintenance history has been for the septic system.

◊ It is *highly* recommended that you obtain all building department permits, surveys, plot plans and approvals for the septic system. The life expectancy of a septic system is about 30 years depending upon the type of construction materials used and the maintenance given to it over the years. The repair costs and costs to re-build or move a septic system are **very high**!!

◊ It is *highly* recommended that you have a licensed septic contractor pump clean and internally inspect the septic system prior to closing. The dye testing performed during a home inspection is **very limited** and does not always reveal a septic system that has failed or is on the verge of failure. When the system is pumped out clean, the septic contractor can *internally* inspect the holding tank and the drainage lines coming into and out of it. This gives him a visual look at the interior of the tank and often the septic contractor will provide a written report for this service. Another benefit for you to get the septic system pumped clean prior to closing, is that if you do buy the house, then they will be moving in with a cleaned out septic tank that should not need any maintenance for quite some time. A septic contractor can also partially dig up the leaching field area to do a more extensive evaluation. This will allow them to determine if the leaching fields or septic drainage pipes are clogged.

◊ Septic systems **MUST** be cleaned at least every 2 to 3 years to properly maintain them. It should be more frequent than every two years if there are a lot of people in the house or if the homeowner does a lot of entertaining and often has guests/parties at the home. It is highly recommended that you obtain all documentation of the past septic maintenance records for future use. Many times a neglected septic system will be pumped clean just because the house is being put on the market for sale. It is similar to driving a car for many years without changing the oil. The car will run on dirty oil, but it will cost you money in wear and tear and it will eventually die prematurely due to the lack of maintenance given to it.

◊ The Owner or Realtor stated that the date the septic system was last pumped clean was _____. They also stated that the septic system was pumped out and cleaned approximately (#)_____ years before this last cleaning.

◊ The septic system was tested by turning on (#)_____ faucets for about (#)_____ minutes. All toilets were flushed (#)_____ times during the testing. A harmless, colored dye designed for septic testing was flushed down the toilets at the beginning of the septic test. The dye did (y/n)_____ show up on or around the property. If the dye is seen or there is a septic odor in the lawn during the testing, then this indicates a problem condition with the septic system and a licensed contractor needs to evaluate the septic system.

_____
_____
_____
_____
_____

Electrical System

◊   The electrical system service entrance line is located _____.
There are (y/n)_____ tree branches touching the electrical lines and equipment. All tree branches near the lines and equipment can cause damage and should be pruned away for safety.

◊   The electrical service entrance head is in (G/A/F/P)_____ condition. There were (y/n)_____ signs of excessive rust or corrosion on the areas that are visible to the inspector.

◊   There was (y/n)_____ a drip loop on the electrical lines before they enter the service entrance head. A *"drip loop"* is created by slack in the wiring in a "U" shape. This helps prevent rainwater from following the electrical lines down into the main panel.

◊   There are (#)_____ service entrance lines going into the house. Two service entrance lines indicate that there is 110 volts inside. A 110 volt electrical system will usually only have a maximum of 30-60 amps of electrical service in the main panel. Three service entrance lines indicate that there is 220 volts inside. A 220 volt electrical system can have up to 200 amp electrical service in the main panel.

◊   The exposed electrical service lines from the service entrance head to the main electrical panel were (y/n)_____ enclosed in a conduit. A *"conduit"* is a covering to protect the electrical lines from the weather and damage. The conduit is in (G/A/F/P)_____ condition and there were (y/n)_____ signs of cracked or open areas and/or joints that are not sealed properly.

◊   The electrical meter is located _____. When the meter is located on the exterior, then the utility company can take a reading without having to enter the house.

◊   **Remember that electricity can kill you!!** Before touching the main panel or any sub-panels check them with a voltage tester to make sure that they are not electrified. Do not go near any exposed wiring or any electrical panels or wiring if there is water on the floor. Water and electricity *don't* mix!!

◊   The main electrical panel is located _____. There were (y/n)_____ signs of excessive rust or corrosion on the main electrical panel. Any sign of excessive rust or corrosion requires a licensed electrician to properly evaluate the electrical system for safety. The main electrical panel is (y/n)_____ installed on the wall securely. There are (y/n)_____ hazardous conditions around the panel, such as, potential water, objects in the way, the panel being too high to reach safely, etc.

◊   There are (y/n)_____ sub-panels noted. *"Sub-panels"* are small electrical panels that branch off from the main electrical panel. The purpose of sub-panels is to prevent very long branch circuit runs in the house.

◊   The electrical system has (circuit breakers/fuses)_____ for the branch circuits. The inspector cannot turn any circuit breakers off or on or replace any fuses, for safety reasons. There are (y/n)_____ tripped circuit breakers or blown fuses in the main electrical panel and/or sub-panels. Any *"tripped"* circuit breakers or *"blown"* fuses indicate a problem that *must* be evaluated by a licensed electrician.

◊   All circuits are (y/n)_____ marked to indicate where each branch circuit leads to. This is a convenience and safety feature. The markings will assist the homeowner in turning off individual branch circuits in case of an emergency or if repairs are needed. There is **no way** for the inspector to determine if the circuits are properly marked for the exact location in the house for their corresponding branch lines.

◊   There are (y/n)_____ open circuit breaker or fuse slots in the main panel and/or sub-panels. Open slots need to be covered with *"blanks"* or spare circuit breakers or fuses. This will prevent anyone from sticking their fingers or any objects inside the electrical panel and getting electrocuted.

◊   There is (y/n)_____ room in the main panel for additional branch circuits. Any unused circuits will generally allow the homeowner to expand the system by adding more branch circuits directly from the main panel without having to install sub-panels.

◊   The main electrical disconnect is located _____. This will shut off **all** of the electrical current leading into the house from the main panel. The main disconnect is (y/n)_____ installed at a safe height that is readily accessible. The main disconnect should be at least 30 inches above the ground and no more than 7 feet high for safety. This will enable it to be safely turned off in case of an emergency.

◊   The amperage for the electrical system is determined to be (#)_____ amps. The National Electrical Code, *(NEC)*, recommends that the minimum amperage be 100 amps for a residential property. The **main disconnect switch** does (y/n)_____ have a visible amperage rating number. The **main electrical panel** does (y/n)_____ have a visible amperage rating number. The **service entrance lines** are (y/n)_____ visible to be viewed for the voltage capacity leading into the house. All three of these indicators must be visible to the inspector to accurately determine the amperage of the electrical system.

◊   The electrical grounding wire is (y/n)_____ installed properly and attached to (water main/grounding rod)_____. There are (y/n)_____ signs of excessive rust or corrosion that require repairs by a licensed electrician. **It is <u>extremely</u> important that the electrical system be grounded to a properly working grounding cable that is attached to the water main line or a grounding rod that extends at least 10 feet into the soil.** The grounding wire and clamps **must** be checked periodically and must not be disconnected or rusty. The purpose of this is to ground the electrical system for safety.

◊   There are (y/n)_____ signs of loose and/or exposed electrical wiring. There are (y/n)_____ signs of loose electrical switches and outlets that need to be secured. Any loose and/or exposed wiring, switches and outlets *must* be secured by a licensed electrician to prevent any electrical hazards.

◊   There are (y/n)_____ signs of *"do-it-yourself"* or unprofessional electrical work. All electrical repairs must be performed by a licensed electrician and all valid permits and building department approvals must be obtained for any work performed.

◊   The outlets and switches are (y/n)_____ reachable from the bathroom tubs or showers. As a safety precaution the outlets and switches in the bathroom **must not** be reachable from the tub or shower. Remember that water and electricity do not mix!!

◊   A limited electrical outlet tester was used to spot check the outlets for proper wiring and current. There are (y/n)_____ outlets noted with improper wiring and/or grounding. Improper wiring can be caused by having the hot and neutral wires reversed in the back of the outlet or a false ground wire. Outlets with these conditions may still provide electrical current but they are an electrical safety hazard and *must* be repaired by a licensed electrician.

◊   There are (y/n)_____ two pronged outlets noted that *must* be upgraded to modern three pronged outlets by a licensed electrician. Older houses will have two pronged outlets as opposed to the modern three pronged type. The third prong is used for the grounding prong in electrical cord plugs. The purpose of this grounding prong is that most appliances today have an internal ground for electrical safety reasons.

◊   There are (y/n)_____ properly operating Ground Fault Circuit Interrupters, *(GFCI)*, noted in some of the outlets and/or in the electrical panel. A *"GFCI"* is an electronic device that will *"trip"* (turn off) the circuit when it senses a potentially hazardous condition. It is very sensitive and operates very quickly. The quick response time in interrupting the power is fast enough to prevent injury to anyone in normal health. GFCI's are recommended by the National Electric Code to be installed anywhere near water for safety. Such water prone areas are basements, garages, kitchens, bathrooms and all exterior outlets.

◊ There are (y/n)_____ an adequate number of electrical outlets noted. There were (y/n)_____ electrical extension cords in use at the time of the inspection. The NEC recommends that houses have an outlet for every 6 feet of horizontal wall space. This is because most appliances come with 6 foot electrical cords and if there are not enough outlets, then the homeowner will have to use extension cords. Extension cord wiring is not recommended because of the possibility of someone plugging an appliance into a low amperage rated extension cord. This will cause the extension cord wire to overheat and start an electrical fire.

◊ It is *highly* recommended that the homeowner install child proof electrical outlet caps if there are any children in the house. These are small plastic plugs to cover any unused outlets so a child will not stick anything into the outlets and get electrocuted.

_____
_____
_____
_____
_____
_____
_____
_____
_____
_____
_____
_____
_____
_____
_____
_____
_____

Additional Comments

The Lower Level Inspection #2

Lower Level

◊   There is (y/n)_____ a lower level area for the subject property.

◊   The lower level is (y/n)_____ finished with areas that are not accessible or not visible. The lower level has (y/n)_____ areas that are inaccessible due to personal items and furniture of the homeowner, or lack of access for the inspector to view areas. Any inaccessible areas cannot be evaluated by the inspector.

◊   The lower level stairs are in (G/A/F/P)_____ condition. The stairs do (y/n)_____ have sturdy handrails with closely spaced posts to prevent tripping hazards and children from falling through the railing openings. The stairs do (y/n)_____ have evenly spaced steps to help prevent tripping hazards.

◊   The house construction materials used for the foundation walls, in the visible areas, are made of _____.

◊   The floor of the lower level does (y/n)_____ have a concrete covering. A concrete floor is recommended in the lower level areas to help prevent moisture and wood destroying insect problems in the house. The floor is in (G/A/F/P)_____ condition where visible and accessible.

◊   There are (y/n)_____ signs of abnormally large settlements cracks in the visible areas of the walls and floors of the lower level. Any **long, horizontal settlement cracks** or any cracks that are over 1/4 of an inch wide **MUST** be further evaluated by a licensed contractor to determine the possible cause and repair options.

◊   There are (y/n)_____ signs of areas of the foundation that have been altered from the time of the original construction of the house. Any alterations to the structure from the time of the original construction will require valid permits and approvals from the municipality for the work performed.

◊   The main girder beam(s) was checked, where accessible, and was in (G/A/F/P)_____ condition and is made of _____. The *"main girder"* of a house is the large beam that spans across the entire width of the house. This is the beam that supports the interior portions of the house and it rests on the top the foundation walls. There are (y/n)_____ signs of excessive rusting, rotted, cut-out, cracked and/or sagging sections of the main girder beam(s). When any of these conditions are noted, the beam(s) **must** be evaluated by a licensed contractor for safety.

◊   The support posts were checked, where accessible, and are in (G/A/F/P)_____ condition and are made of _____. The *"support posts"* of a house are found underneath the main girder beam(s) and should be spaced about 6 feet apart. These posts support the main girder beam(s) in the middle sections of the house lower level area while the ends of the main girder beam(s) are supported by the foundation walls. There are (y/n)_____ signs of excessive rusting, rotted, cut-out, cracked and/or sagging sections of the support posts. When any of these conditions are noted, the posts **must** be evaluated by a licensed contractor for safety.

◊   The floor joists were checked, where accessible, and are in (G/A/F/P)_____ condition. The *"floor joists"* are the wood boards that span across the underside of the floors in the house, which in turn hold up the floors. The floor joists run perpendicular to the main girder beam(s). The floor joists do (y/n)_____ have the required diagonal bracing installed in the visible areas. These are small wood boards or metal straps placed diagonally in between the floor joists to *"tie"* them together for additional strength. There are (y/n)_____ signs of excessive rotted, cut-out, cracked and/or sagging sections of the floor joists. When any of these conditions are noted, the joists **must** be evaluated by a licensed contractor for safety.

◊   The sub-flooring was checked, where accessible, and was in (G/A/F/P)_____ condition. The *"sub-flooring"* is the plywood boards or paneled wood boards that are located on top of the floor joists. The purpose of the sub-flooring is to support the finished flooring above, such as hardwood, tiles or carpeting, that rests on top of the sub-flooring. There are (y/n)_____ signs of excessive rotted, cut-out, cracked and/or sagging sections of the sub-flooring.

_____
_____
_____
_____
_____
_____
_____
_____
_____
_____
_____
_____
_____
_____
_____
_____
_____
_____
_____

Crawl Spaces

◊ There is (y/n)_____ a lower level crawl space area for the subject property.

◊ The inspector did (y/n)_____ have adequate access to view the crawl space area. *"Crawl spaces"* are small areas underneath the livable portions of the house which are not high enough to stand up in. This is an area that **demands** attention periodically since there is a higher risk of rot and termite infestation due to it being dark and damp most of the time.

◊ There does (y/n)_____ appear to be adequate ventilation in the crawl space area. Crawl spaces need plenty of ventilation all year round to help prevent rot and wood destroying insect infestation.

◊ The crawl space floor has a (dirt/concrete/etc)_____ covering. Any crawl spaces that have a dirt floor should be covered with a concrete surface. Dirt floors will promote moisture from the soil and are an attraction to wood destroying insects. If putting concrete over the dirt floor is too expensive then a 6 mil plastic floor cover can be placed over the dirt areas to help eliminate some of the moisture problems in the crawl space.

◊ There are (y/n)_____ signs of abnormally large settlements cracks in the visible areas of the walls and floors of the crawl space. Any **long, horizontal settlement cracks** or any cracks that are over 1/4 of an inch wide **MUST** be further evaluated by a licensed contractor to determine the possible cause and repair options.

◊ There are (y/n)_____ signs of areas of the foundation in the crawl space area that have been altered from the time of the original construction of the house. Any alterations to the structure from the time of the original construction will require valid permits and approvals from the local municipality for the work performed.

◊ The main girder beam(s) of the crawl space was checked, where accessible, and was in (G/A/F/P)_____ condition and is made of _____. There are (y/n)_____ signs of excessive rusting, rotted, cut-out, cracked and/or sagging sections of the main girder beam(s). When any of these conditions are noted, the beam(s) **must** be evaluated by a licensed contractor for safety.

◊ The support posts of the crawl space were checked, where accessible, and are in (G/A/F/P)_____ condition and are made of _____. There are (y/n)_____ signs of excessive rusting, rotted, cut-out, cracked and/or sagging sections of the support posts. When any of these conditions are noted, the posts **must** be evaluated by a licensed contractor for safety.

◊ The floor joists of the crawl space were checked, where accessible, and are in (G/A/F/P)_____ condition. The floor joists do (y/n)_____ have the required diagonal bracing installed in the visible areas. There are (y/n)_____ signs of excessive rotted, cut-out, cracked and/or sagging sections of the floor joists. When any of these conditions are noted, the joists **must** be evaluated by a licensed contractor for safety.

◊ The sub-flooring of the crawl space was checked, where accessible, and was in (G/A/F/P)_____ condition. There are (y/n)_____ signs of rotted, cut-out, cracked and/or sagging sections of the sub-flooring.

_____

_____

_____

_____

Gas Service

◊ The subject property reportedly is (y/n)_____ connected to local gas utility lines in the street. When there is no gas service in the building, you should determine from the local utility company whether or not it is available and what costs are involved in having gas service installed in the house. Installing gas service can be a major expense to the homeowner.

◊ The gas meter is located _____. The gas meter and visible gas lines are in (G/A/F/P)_____ condition. There are (y/n)_____ signs of excessive, loose and/or leaking gas lines. The visible gas lines are made of _____ and do (y/n)_____ appear to be an approved type of gas piping. All gas service lines should be approved black iron gas piping. Any loose, leaking, excessive rusting conditions, or unapproved types of metals being used for gas feed lines **must** be evaluated by a licensed contractor to make any necessary repairs to bring the gas lines up to the building codes.

◊ There is (y/n)_____ a main shut-off valve near the gas meter for safety. This is used to shut-off the gas supply for repairs or in case of an emergency. If you smell or detect any gas leaks in the house, **immediately** contact the local utility company to make repairs. **Leaking gas will explode so *do not* take any chances!!**

◊ There is (y/n)_____ Liquid Petroleum Gas *(LPG)* tank(s) noted on the site at the time of the inspection. The LPG gas tank(s) is in (G/A/F/P)_____ condition. There are (y/n)_____ signs of excessive rust or corrosion. The tank(s) does (y/n)_____ appear to be properly level on a sturdy foundation. Any problem condition **must** be checked out a reputable LPG contractor for safety. It is recommended that you check with town hall to make sure all valid permits and approvals are on file for the existence of any LPG tank(s) on the site.

◊ **Do not** bring any gas tanks into the house, such as, exterior barbecue tanks or automobile gas cans. Barbecue gas tanks are under extreme pressure, similar to scuba diving tanks. *If they ever exploded they could blow up the entire building and everyone in it!!*

_____

_____

_____

_____

_____

_____

_____

_____

_____

_____

Auxiliary Systems

◊ There is (y/n)_____ an auxiliary system(s) for the subject property.

◊ Auxiliary systems, such as, burglar alarm systems, fire detection systems, intercoms, central vacuum systems, lawn sprinklers, etc., *are not* evaluated during a limited time home inspection. Obtain all manuals from the Owner and find out how to operate these systems, when they are present on the site. Determine from the Owner or Realtor if any fire or alarm systems are hooked up to any monitoring services and/or the local police or fire departments and what the fees are for this service.

◊ The following auxiliary systems were present on the site:

_____

_____

_____

_____

Water Penetration

◊ Signs of water penetration can be white mineral salts on the concrete walls and floors. This is called *"efflorescence"* and is caused by water seeping through the concrete and then drying on the exterior portion leaving the white, mineral salt from the cement as a residue.

◊ Most lower level areas will get some minor efflorescence on the lower portion of the walls and floors due to normal humidity in this area because it is underground. It is recommended that you use a dehumidifier to help prevent moisture in any lower level areas.

◊ Sometimes in the corners there may be indications of water stains. Often this is caused by the lack of gutters and downspouts on the house or downspouts that drain right next to the foundation walls on the exterior. All downspouts should be piped at least 5 feet away from the house so the rainwater will not drain next to the foundation and then enter the lower level of the house. Sometimes the downspouts drain into underground drainage lines. These lines can become clogged due to leaves or a small animal getting stuck in them. Gutters, downspouts and underground drainage lines need to be checked periodically for proper operation.

◊ The grading of the soil next to the exterior of the house can also cause water stains on the lower level walls and floors. All soil next to the foundation should be slightly sloped away from the side of the house to help prevent rainwater from entering the lower level.

◊ There are (y/n)_____ signs of abnormal or excessive water problems in the house beyond normal humidity and condensation stains.

◊ The accessible wood members that are in contact with the floor, such as, workbench posts, storage items, etc., were probed to reveal any rotting conditions or evidence of abnormal or excessive water penetration. The wood members that were spot checked are in (G/A/F/P)_____ condition and did (y/n)_____ have signs of abnormal or excessive water penetration.

◊   There is (y/n)_____ a sump pump located in the lower level. These are small pumps that help carry water away from the house. Sump pumps are located in small pits dug into the lower level floor and have a drainage pipe to carry water to a more desirable location.

◊   The sump pit walls are in (G/A/F/P)_____ condition. Any *"do-it-yourself"* or other unprofessional installations need to be repaired by a licensed contractor. There was (y/n)_____ water inside the sump pit at the time of the inspection. Any water in the sump pit indicates that the area has a high groundwater table and that there is a potential for water to enter the lower level of the house.

◊   The sump pump is in (G/A/F/P)_____ condition. The pump did (y/n)_____ operate properly when tested at the time of the inspection. The sump pump is (y/n)_____ plugged into an outlet with a GFCI for safety. The sump drainage line does (y/n)_____ have a required backflow preventer. This is a check valve inside one section of the line to help prevent the drainage water from flowing backwards towards the sump pit after it is pumped out. The sump drainage line does (y/n)_____ discharge the water at least 5 feet away from the exterior foundation of the house. This is required to prevent the water from flowing back into the sump pit after it has already been pumped out from the lower level of the house.

◊   It is *highly* recommended that you check the local building department to determine if the subject property is located in a designated flood hazard zone. A *"flood hazard zone"* is an area where the government has determined that there is a potential of the area becoming flooded from time to time. Flood maps are located in every town hall and are available to the public to view. If a house is located in a flood hazard zone, then the homeowner should obtain flood hazard insurance in addition to normal homeowner and title insurance policies.

_____
_____
_____
_____
_____
_____
_____
_____
_____
_____

Additional Comments

The Interior Home Inspection #2

Kitchen

◊ The kitchen walls and floors are in (G/A/F/P)_____ condition. There are (y/n) _____ signs of structural problems or abnormal settlement cracks that need further evaluation by a licensed contractor.

◊ The kitchen floor covering is in (G/A/F/P)_____ condition. The floor covering is made of _____.

◊ The kitchen cabinets are in (G/A/F/P)_____ condition. The cabinets are (y/n)_____ securely fastened to the walls and/or floor. It is recommended that the homeowner install child guards on the cabinets and drawers if there are any children in the home.

◊ The kitchen countertop is in (G/A/F/P)_____ condition. The countertop is (y/n)_____ securely fastened.

◊ The kitchen faucet was tested and there was (y/n)_____ adequate hot water. There were (y/n)_____ leaks in the faucet and/or underneath the sink. There is (y/n)_____ a properly operating spray attachment in the sink area.

◊ There was (y/n)_____ water filter device connected to the kitchen sink water lines. A water filter device is *highly* recommended for health reasons.

◊ There are (y/n)_____ an adequate number of outlets for modern usage in the kitchen. The outlets do (y/n)_____ have the required three prongs with GFCI protection for safety.

◊ The Owner or Realtor stated that the appliances are (y/n)_____ being sold with the house. The appliances were (y/n)_____ spot checked for proper operation. The inspector *cannot* properly evaluate appliances during a limited time inspection. All owners manuals and operating instructions should be obtained from the Owner.

_____

_____

_____

_____

_____

Bathrooms

◊   There is (#)_____ bathroom(s) in the house.

◊   The bathroom walls and floors are in (G/A/F/P)_____ condition. There are (y/n)_____ signs of structural problems or abnormal settlement cracks that need further evaluation by a licensed contractor.

◊   The bathroom floor covering is in (G/A/F/P)_____ condition. The floor covering is made of _____.

◊   The bath tub and shower area did (y/n)_____ have loose or cracked sections. This area was (y/n)_____ in need of grout/caulk at the time of the inspection. Grout is needed to prevent water leaks behind the walls and floors.

◊   The bathroom cabinets are in (G/A/F/P)_____ condition. The cabinets are (y/n)_____ securely fastened to the walls and/or floor. It is recommended that the homeowner install child guards on the cabinets and drawers if there are any children in the home.

◊   The bathroom sink top(s) is in (G/A/F/P)_____ condition. The sink top(s) is (y/n)_____ securely fastened.

◊   The bathroom sink, tub and/or shower faucets were tested and there was (y/n)_____ adequate hot water. There were (y/n)_____ leaks in the faucets and/or underneath the sink(s). The drain stop mechanism(s) in the sink(s) and bath tub(s) are (y/n)_____ working properly.

◊   The water pressure and drainage noted during the testing was (y/n)_____ normal and adequate with no problems indicated. The client was (y/n)_____ present during the testing.

◊   There are (y/n)_____ an adequate number of outlets for modern usage in the bathroom(s). The outlets do (y/n)_____ have the required three prongs with GFCI protection for safety.

◊   Any Jacuzzi or hot tub devices *cannot* be evaluated during a limited time inspection. All owners manuals and operating instructions should be obtained from the Owner.

_____
_____
_____
_____
_____
_____
_____
_____
_____

Floors and Stairs

◊ The floors are (y/n)_____ in sound structural condition. There are (y/n)_____ signs of sagging or uneven sections due to abnormal structural settlement.

◊ There are (y/n)_____ finished hardwood floors noted in the rooms. The hardwood floors are in (G/A/F/P)_____ condition where visible.

◊ There are (y/n)_____ carpets noted in the rooms. The carpets are in (G/A/F/P)_____ condition. The inspector was (y/n)_____ able to check underneath the corner of some of the carpeting and/or in the closets to determine what type of flooring is underneath. _____ flooring was noted underneath the carpets and/or in the closets.

◊ If the Owner has any pets it is recommended that all carpets be fumigated or removed prior to taking possession of the house. This will prevent the possibility of moving-in and finding fleas in the carpeting.

◊ All staircases are (y/n)_____ sturdy and the steps do (y/n)_____ have even and safe stair heights to help prevent tripping hazards. All stairs over two steps in height do (y/n)_____ have secure handrails for safety. All handrails do (y/n)_____ have closely spaced posts to help prevent children from falling through the railings. There are (y/n)_____ light fixtures and light switches at the top and bottom of all stairways for safety.

◊ All windows at the base of staircases *must* have a sill height of at least 36 inches above the floor. This will help prevent someone from falling through the window in the event of a fall down the stairs. If the sill is less than 36 inches high, a window guard **must** be installed as a precautionary measure.

_____
_____
_____
_____

Walls and Ceilings

◊ The walls and ceilings are (y/n)_____ in sound structural condition. There are (y/n)_____ signs of abnormal settlement cracks or structural problems that need further evaluation by a licensed contractor. There are (y/n)_____ minor settlement cracks noted in the walls and/or ceilings that appear to be from the normal expansion and contraction of the construction materials.

◊ The interior walls are made of (sheetrock/plaster/etc)_____.

◊ There are (y/n)_____ signs of water stains or water damage in the visible areas of the walls and/or ceilings. Any water stained areas will have more extensive damage behind the walls and ceilings that is not visible to the inspector and *must* be evaluated by a reputable contractor. The finished coverings will have to be removed to expose this hidden area in order to view the damage.

◊   The interior walls and ceilings do (y/n)_____ need to be painted at this time. All settlement cracks and damaged areas should be patched and sealed over the next time the interior is re-painted. The inspector *cannot* determine if there is any lead paint in the house. This can only be determined by a reputable laboratory analysis which should be performed prior to closing if any doubts exist.

◊   There is (y/n)_____ wallpaper noted on the interior walls. The wallpaper is in (G/A/F/P)_____ condition and there were (y/n)_____ signs of peeling or aging sections. Removing any existing wallpaper is a time consuming job that can be expensive. All cost estimates should be obtained. You have to be careful removing wallpaper from sheetrock walls because you can pull the cardboard paper covering off the sheetrock walls along with the wallpaper.

◊   There are (y/n)_____ smoke detectors noted on all levels in the home. The smoke detectors were (y/n)_____ able to be spot checked for proper operation. The smoke detectors do (y/n)_____ appear to be in operating order. The smoke detectors are operated by (batteries/electricity)_____. Smoke detectors are **required** on all levels of a residence and heat detectors are recommended in the garage, attic and lower level areas. If the smoke detectors are battery operated it is *highly* recommended hat you replace all batteries after moving in. If the smoke detectors have a *"hard wired"* installation, obtain all operating instructions from the homeowner. Smoke detectors **cannot** be properly evaluated during a limited time home inspection.

_____

_____

_____

_____

Windows and Doors

◊   The windows are in (G/A/F/P)_____ condition and are (y/n)_____ operating properly. The doors are in (G/A/F/P)_____ condition and are (y/n)_____ operating properly. The windows and doors were spot checked by opening and closing at least one window and door in each room.

◊   There were (y/n)_____ cracked or broken panes of glass noted at the time of the inspection. There were (y/n)_____ broken vacuum seals noted in the thermal windows and/or doors. When this seal is broken, it allows moisture to get trapped between the two window panes of thermal glass and leaves condensation stains. Broken vacuum seals can sometimes be repaired by a glass service company but estimates should be obtained prior to closing.

◊   The window locks are (y/n)_____ operating properly. The door locks are (y/n)_____ operating properly. The locks were spot checked by the inspector. There are (y/n)_____ double key locks noted on the doors. *"Double key"* locks require a key to exit and enter through the door. The purpose of these locks is so that if a burglar breaks a door window, they cannot just turn a bolt and open the door, instead they will need a key to open the lock. **However, in some areas these locks are against the local fire codes because a key is needed to exit in the event of an emergency which can cause people to get trapped inside the house during a fire.** Check with the local fire and building departments for their recommendations about door locks. All locks *must* be replaced by a licensed locksmith upon taking possession of the house for security reasons.

The Interior Home Inspection #2 www.nemmar.com   145

_____
_____
_____
_____
_____
_____
_____

Fireplaces

◊   There is (#)_____ fireplace(s) in the house.

◊   The fireplace(s) has (y/n)_____ signs of structural problems. The mortar joints and/or firebox area are in (G/A/F/P)_____ condition.

◊   The fireplace(s) has (y/n)_____ signs of back-smoking problems. Back-smoking is caused by downdrafts in the chimney flue which cause the smoke to come back into the house. Signs of back-smoking are black deposits, called *"creosote,"* on the front of the fireplace and mantel.

◊   The fireplace(s) does (y/n)_____ have a properly operating sliding screen cover and glass doors for safe and energy efficient operation of the fireplace(s).

◊   The fireplace damper(s) is in (G/A/F/P)_____ condition and is (y/n)_____ operating properly. The *"damper"* is the metal door inside the top of the firebox area which is opened while a fire is burning and closed when the fireplace is not in use. If the damper does not open and close properly or is very rusty it must be replaced.

◊   The inspector was (y/n)_____ able to view inside the chimney flue(s). The flue lining(s) is made of (brick/tile/cement/etc)_____. The visible areas of the flue lining(s) are in (G/A/F/P)_____ condition. There are (y/n)_____ signs of an excessive buildup of creosote inside the flue lining(s). All chimney flues need to be swept and repointed by a reputable chimney sweep periodically to help prevent chimney fires from thick creosote deposits.

_____
_____
_____
_____

Attic Inspection

◊   The attic space was (y/n)_____ accessible for the inspector to make evaluations of the areas above the top floor finished ceilings.

◊   The access panel leading to the attic is located _____.
There is (y/n)_____ a stairway to provide easy access to the attic. The stairway was in (G/A/F/P)_____ condition and did (y/n)_____ have sturdy handrails and steps for safe passage.

◊   It is *highly* recommended that a handrail be installed inside the attic area surrounding the access opening. This will help prevent anyone walking in the attic area from falling through the access opening. Do not wait for an accident to happen, install a handrail now!!

◊   The attic floor is (y/n)_____ covered so that this area can be used to store **lightweight** household items. Heavy objects are not recommended for storage in the attic area due to the excessive weight they exert on the ceiling below.

◊   The roof ridge beam appeared to be in (G/A/F/P)_____ condition where visible and accessible. The roof *"ridge beam"* is the main girder type beam at the top of the crest of the roof.

◊   The roof rafters appeared to be in (G/A/F/P)_____ condition where visible and accessible. The roof *"rafters"* are the floor joist type beams leading from the attic floor up to the roof ridge beam.

◊   The roof sheathing appeared to be in (G/A/F/P)_____ condition where visible and accessible. The roof *"sheathing"* is the sub-flooring type wood that the roof shingles rest upon. The sheathing is made of (plywood/etc)_____.

◊   There were (y/n)_____ water stains that appeared to be due to water leaks or abnormal humidity in the attic area. Often there are old water stains from prior roof leaks that have been repaired. There are (y/n)_____ bowed and/or damaged sections of the wood members in the attic roof area. Any water leaks, abnormal humidity, bowing and/or damaged wood members in the attic roof area will require repairs by a licensed roofing contractor.

◊   There are (y/n)_____ collar beams in the attic. *"Collar beams"* are generally constructed of 2 x 4 inch wood boards that are located several feet below the ridge beam. The purpose of collar beams is to *"tie"* both sides of the roof together so that all of the weight of the roof does not rest upon the ridge beam alone. Collar beams give the roof additional support and it is recommended that they be installed when not noted.

_____
_____
_____
_____
_____
_____

Attic Ventilation

◊ It is very important that the attic be properly ventilated to help prevent excessive humidity or heat in this area. Even in cold winter months humidity can cause problems in attics and needs to be ventilated to the exterior. In the summer months attics can reach 150 degrees Fahrenheit, which adds a big heat load on the house.

◊ There size of the attic vent(s) do (y/n)_____ appear to be adequate to provide proper ventilation in this area.

◊ The screens on the attic vents are in (G/A/F/P)_____ condition. The screens need to be kept clean at all times. Screens help to keep birds and bees out of the attic.

◊ There were (y/n)_____ bathroom fans noted to be discharging in the attic area. All bathroom vents and fans *must* discharge to the exterior of the house. If they discharge in the attic they will create moisture problems.

◊ There are (y/n)_____ soffit vents. *"Soffit vents"* are vents at the base of the roof where it overhangs the exterior siding. Soffit vents are recommended to allow air to come into the base of the roof in the attic area and carry any unwanted heat or moisture out the attic gable or roof vents.

◊ There is (y/n)_____ a ridge vent. A *"ridge vent"* is a vent at the very top of the roof, above the ridge beam. Ridge vents are recommended to allow unwanted heat and moisture to escape from the attic through the top of the roof.

◊ There is (y/n)_____ a thermostatically operated power ventilator in the attic. A *"thermostatically operated power ventilator"* is a fan in the roof that operates by a thermostat. When the temperature in the attic reaches a pre-set level, the fan will automatically turn on to cool this area. When the temperature drops low enough, the fan will automatically turn off by itself.

◊ There is (y/n)_____ an attic and/or whole house fan installed in the house. Attic and house fans need to have adequate vents to discharge the air so the fan will operate properly and will not break down prematurely.

_____
_____
_____
_____
_____
_____
_____
_____
_____
_____

Attic Insulation

◊   There is (y/n)_____ insulation noted in the floor joists of the attic area where visible and accessible. Insulation in the **attic floor joists** is required for energy efficiency in a house. Insulation is *not* needed in between the **roof rafters** because once heat has escaped through the upper level ceiling it is lost anyway. There is no sense trying to trap it in the attic, you will only be trapping unwanted moisture along with the heat in the attic if you install insulation in the roof rafters.

◊   The insulation is approximately (#)_____ inches thick. It should be at least 8 inches thick for energy efficiency. Installing additional insulation will increase energy efficiency in the house. Any air-conditioning or heating ducts in the attic must be insulated either on the exterior or the interior for energy efficiency.

◊   The type is insulation noted is made of _____. The vapor barrier is (y/n)_____ installed properly. The *"vapor barrier"* is the aluminum foil layer on one side of the insulation roll. It must always be touching the **heated** side of the building. For example, if it is installed in an attic the vapor barrier must face downwards. If it is installed in an unheated basement or crawl space the vapor barrier must face upwards. The reason for the vapor barrier is to prevent any moisture from getting trapped in between the insulation and condensing into water, which will decrease the energy efficiency. It prevents moisture problems by reflecting the heated air, which has moisture in it, back towards the heated portion of the house. If an existing layer of insulation has a vapor barrier, then if you add more insulation on top of the existing layer, the new layer added should **not** have a vapor barrier since the aluminum foil barrier will trap moisture in between both layers of insulation.

◊   The Owner or Realtor stated that they have (y/n)_____ installed insulation into areas of the structure. They stated that they do (y/n)_____ know if any prior owner's have installed insulation in the structure. Any *"blown-in"* type of insulation can have substances in it that are a health hazard. In the past, some houses had UFFI insulation blown into the walls and floors. UFFI stands for *Urea Formaldehyde Foam Insulation* and the Environmental Protection Agency, or *EPA*, has issued warnings about this type of insulation. If there is any UFFI or other type of unknown insulation in the house, it is **highly** recommended that an air sample be taken by a licensed laboratory to see if there are any health concerns with this insulation in the house.

_____
_____
_____
_____
_____
_____
_____
_____

Asbestos Insulation

◊ Many older houses will have asbestos insulation on the heating pipes. Sometimes the old cast iron boilers have asbestos on the interior insulating walls also. Believe it or not, when asbestos first came out it was <u>required</u> to be installed in all new construction. That is why so many buildings have it. It was considered a "miracle product" when it first came out because Asbestos has great insulating and fireproofing qualities. The only problem was that they did not know about the health problems associated with it until it was too late.

◊ Asbestos causes lung cancer when it comes loose from the pipes and the fibers get into the air. Asbestos fibers are like tiny daggers and when you breathe them in, they stick into your lungs and stay there. The fibers cling to dust and can be stirred up off of the floor when someone walks in a basement. There are about 5 different diseases that are related to exposure to asbestos.

◊ The Environmental Protection Agency, *(EPA),* has offices in every State which will provide anyone with free information and brochures on Asbestos, Radon Gas, Fuel Oil Leaks, Water Quality, and a lot of other environmental and health concerns that the homeowner needs to know about. Call your State EPA office and obtain their brochures for more information and advice.

◊ There is (y/n)_____ evidence of possible Asbestos insulation in the house. Asbestos insulation usually has a white color and appears to have layers of ribbed cardboard in the middle sections. It is usually wrapped in an off-white canvas covering. However, the only way to know for sure if any insulation is Asbestos is to have a laboratory take an air sample. **It is recommended that you contact a licensed Asbestos contractor or laboratory whenever there is <u>any</u> evidence of possible Asbestos insulation or if any doubts exist.**

◊ There are (y/n)_____ signs of residue or loose sections of Asbestos insulation in the house. The Environmental Protection Agency recommends that Asbestos insulation be <u>**PROFESSIONALLY**</u> sealed or removed from the residence by an EPA *licensed* Asbestos contractor. This means, that the homeowner, the plumber, heating contractor, or any other handyman **SHOULD NOT TOUCH** any Asbestos in the house. Residue from Asbestos insulation on the heating pipes is sometimes noted by the existence of small white particles on sections of the pipes. **This indicates that an Non-EPA licensed person removed the Asbestos!! It is <u>highly</u> recommended that an EPA licensed Asbestos contractor or laboratory be contacted for further evaluations prior to closing on the property if any evidence or doubts exist.**

◊ When an EPA licensed contractor removes Asbestos, they seal off the entire area where it is located and work with completely sealed suits over their bodies. They then set up a vacuum to remove **all** of the dust from the area. When the Asbestos is totally removed from the house, they then take an air sample to make sure they have not left any fibers lying around to be stirred up and breathed in later. Generally, any Asbestos behind the walls is left alone. If there is no access to it and it cannot be disturbed, there usually is not much of a health concern.

_____
_____
_____
_____
_____
_____

Radon Gas

◊   Radon is a radiation gas that is released naturally by rocks and soil in the earth. It gradually seeps up from the ground, and as long as it goes out into the open air it is not a problem. However, if the radon seeps through cracks in the foundation floor and walls it will become trapped in the house and the radon levels will rise. As with asbestos and other environmental and health concerns, call your State Environmental Protection Agency office for their information, brochures and advice. **EPA considers radon to be the number 2 leading cause of lung cancer behind smoking, so it is not something to take too lightly.**

◊   EPA uses a reading of 4 Pico Curies per liter to determine the maximum radon level in a house before mitigation is recommended. Just to give you an idea of how Pico Curies are measured, EPA says that 1 Pico Curie is the average indoor radon level and it is equal to getting about 100 chest X-rays per year. Now that may seem very high, but to put it in the proper perspective, EPA also says that the amount of radiation you receive from a normal chest X-ray, usually is not as high as most people think. For example, with a reading of 1 Pico Curies per liter, EPA estimates that 3-13 people out of 1,000 will die from lung cancer. This is similar to a non-smokers risk of dying from lung cancer. With a reading of 4 Pico Curies per liter, it is estimated that 13-50 people out of 1,000 will die from lung cancer. This is similar to 5 times the non-smokers risk of dying from lung cancer. The lung cancer risk increases as the radon levels and time of radon exposure increases.

◊   *"Mitigation"* is the term used to reduce the problem by lowering the radon levels. When a house is mitigated, the radon contractor will seal all open cracks in the lower level walls and floors. They then drill a hole in the foundation floor, which will look like a sump pump pit. Instead of installing a sump pump in this pit, they install a fan with pipes leading to the outside of the house. In some areas, the local codes require that these pipes discharge __above__ the roof line to prevent the radon from entering back into any open windows on the side of the house. The purpose of the mitigation treatment is to vent the radon gas that builds up underneath the foundation, to the exterior of the house.

◊   In some areas the radon levels tend to be higher than in other areas but **ALL HOUSES WILL HAVE SOME RADON!!** Even if it is minor trace element readings of 0.5 Pico Curies per liter, which can be a minimal health risk. However, you might not have a high radon reading today but you might have a high reading a month from now. Or you might have a high reading and your neighbor might not. The reason for this is that radon is a radiation gas that is unstable and the levels fluctuate often. There are a lot of factors that effect the radon level in a house, such as: 1) The time of the year and the climate. 2) The type of soil and rocky terrain in the area. 3) The type of construction of the building. 4) And there are other factors also. This is why EPA recommends that you re-test for radon every 6 months to make sure that the levels in your house are acceptable. Believe it or not, radon can even be found in water.

◊   Radon testing is usually done with a small, round metal canister with charcoal inside. The canister is left in the house for about 3-5 days and then it is sealed and mailed back to the radon lab for analysis. Generally, all the canister does is absorb the air in the room where it is placed so the lab can analyze the radon levels. The canister does not present a health risk to the occupants of the home. Radon testing canisters *must* be purchased from a licensed EPA laboratory with sophisticated analyzing equipment. **Do not just buy radon cans off the shelf of the local hardware store. The reason for this is that what makes a radon reading accurate is not the canister you use, but the sophistication of the labs analyzing equipment. You could send the same canister to two different labs and get two totally different radon level readings.**

◊   Most labs recommend that you place the canister about 3 feet above the floor in the lowest area of the house. The Environmental Protection Agency feels that if a basement has the potential to become a livable area in the future, then that is where the radon levels should be measured. The basement is where you are going to get the highest radon reading in the home. Any readings on the first floor will generally be lower then basement readings.

Additional Comments

The Exterior Home Inspection #2

Roof

◊   There are (y/n)_____ tree branches overhanging the roof area. There should not be any tree branches overhanging the roof, which can cause damage to the shingles.

◊   The inspector was (y/n)_____ able to view the roof adequately from the following location(s) _____. All judgments and evaluations are made from that perspective.

◊   There are (y/n)_____ bowing sections of the roof ridge beam, roof rafters and/or the roof sheathing which would indicate repairs being needed.

◊   The roof covering/shingles installed on the house is made of _____ _____. The roof covering/shingles appears to be approximately (#)_____ years old. It is in (G/A/F/P)_____ condition. There are (y/n)_____ signs of rapidly aging and damaged areas of the roof covering.

◊   The average life expectancy of this type of roof covering/shingles is (#)_____ years. The life expectancy of all roofs depends upon many factors, some of which are the quality of the roof covering, the quality of the installation, the climate and exposure to the elements and the maintenance given to the roof over the years.

◊   The slope of the roof that has a southerly or southwesterly exposure faces the sun more often and can become brittle and show signs of aging faster. The slope of the roof that has a northerly or northeastern exposure is more apt to have mold and decay fungi due to the lack of sunlight.

◊   There appears to be (#)_____ layer(s) of shingles on the roof. Some local building codes allow there to be up to 3 layers of shingles on a roof. However, this adds too much weight to the structure and it is only recommended that there be 2 layers as a maximum. Also, when shingles are placed over an existing layer they tend to have a poor cosmetic appearance and it cuts down their life expectancy. When there are 2 or more layers of roof covering installed presently, then you will have to strip these layers off during the next re-roofing. This is much more expensive than just having a new layer put over the existing shingles. *Especially* if the roof sheathing has to be replaced as well.

_____
_____
_____
_____

Chimney

◊ There is (y/n)_____ a chimney for the subject property.

◊ The chimney(s) is (y/n)_____ leaning. Any leaning conditions indicate a serious structural problem where they chimney may need to be rebuilt.

◊ The chimney(s) is in (G/A/F/P)_____ condition. There were (y/n)_____ signs of maintenance and repairs being needed due to the exposure to the elements.

◊ The mortar joints are in (G/A/F/P)_____ condition. There is (y/n)_____ evidence that they need to be re-pointed. *"Re-pointing"* refers to putting more cement in the joints mortar joints. The mortar joints between the construction materials need to be checked periodically for deterioration problems to help prevent water penetration.

◊ The chimney(s) is (y/n)_____ made of metal materials. The metal chimney materials noted are in (G/A/F/P)_____ condition. Metal chimneys need to be checked periodically for rust and water leaks.

◊ There is (y/n)_____ an antenna attached to the chimney or the roof. The connections of an antenna must be kept caulked so that water will not enter the house. If there is cable TV in the house and the antenna is no longer in use, then it should be removed. Antennas add stress to the roof and chimney when they move around in the wind, which can create water leaks.

◊ The chimney(s) lining was (y/n)_____ visible from vantage points on the subject property. There is (y/n)_____ a weather cover noted above the top of the chimney stack to prevent water from entering the flue. There is (y/n)_____ a screen noted over the top of the flue stack to prevent animals, such as squirrels, raccoons and birds, from entering the flue.

_____

_____

_____

_____

Siding

◊ The siding on a house is used to provide weather protection. The siding does not support the building structurally, unlike a load bearing wall which does give structural support to the house.

◊ The siding installed on the house is made of _____
_____. It is in (G/A/F/P)_____ condition. There are (y/n)_____ signs of rapidly aging and damaged areas of the siding.

◊ The average life expectancy of this type of siding is (#)_____ years. The life expectancy of all siding depends upon many factors, some of which are the quality of the siding, the quality of the installation, the climate and exposure to the elements and the maintenance given to the siding over the years.

◊ The siding does (y/n)_____ need to be painted or stained at this time. Painted wood will have a more uniform appearance but needs more maintenance. Stained wood will have spotty areas due to the wood absorbing the stain unevenly in some sections. However, staining wood will last a lot longer than painting wood.

◊ All joints around windows and doors do (y/n)_____ appear to be caulked properly to help prevent water penetration and increase energy efficiency.

◊ All areas of the siding are (y/n)_____ at least 8 inches above the soil all around the structure. This is required to help prevent termite and rot problems.

_____
_____
_____
_____
_____
_____
_____
_____

Fascia, Soffits and Eaves

◊ The fascia, soffits and eaves are the molding areas at the bottom of the roof and the top of the siding. It is the small area where the roof overhangs the sides of the house.

◊ The fascia, soffits and eaves are in (G/A/F/P)_____ condition. There are (y/n)_____ signs of rapidly aging and damaged areas at this time. They are (y/n)_____ in need of painting at this time.

◊ There are (y/n)_____ vents noted at the bottom of the roof overhang area. When vents are noted, it indicates that the attic area may have soffit vents. *"Soffit vents"* allow air to enter the bottom of the attic area to help to remove any unwanted heat and moisture from the house attic.

_____
_____
_____
_____

Gutters, Downspouts and Leaders

◊   **Gutters** are installed along the bottom edge of the roof to catch the rainwater running off the roof. **Downspouts** are installed near the ends of the gutters and are used to drain the water from the gutters. **Leaders** are installed at the bottom of the downspouts to direct the rainwater away from the side of the house.

◊   The gutters, downspouts and leaders are made of _____. They are in (G/A/F/P)_____ condition. There is (y/n)_____ evidence of loose or leaning sections that need to be secured. The gutters, downspouts and leaders do (y/n)_____ appear to be clogged with leaves and/or twigs at this time.

◊   There are (y/n)_____ an adequate number of gutters, downspouts and leaders on the house. There should be at least 1 downspout for every 30 feet of gutter to help prevent excessive weight damaging the gutters due to the rainwater.

◊   All downspouts do (y/n)_____ have leaders to pipe the rainwater at least 5 feet away from the foundation to help prevent water problems in the lower level areas.

◊   There are (y/n)_____ downspouts that drain directly into the ground. These generally lead to dry wells or underground drainage lines. At the time of the inspection, they did (y/n)_____ appear to be clogged. They need to be checked periodically for clogging due to leaves and small animals getting stuck in them.

_____

_____

_____

_____

Windows, Screens and Storms

◊   The exterior window frames are in (G/A/F/P)_____ condition. There is (y/n)_____ evidence of rot and/or damaged areas. They do (y/n)_____ need to be painted or stained at this time.

◊   The storm windows and/or screens over the windows are in (G/A/F/P)_____ condition. Storm windows are *highly* recommended in colder climate areas. More heat is lost in a house through the windows than through any other area. Storm or thermal windows can reduce the heat loss by as much as 50%.

_____

_____

_____

Entrances, Steps and Porches

◊ All accessible entrances, steps and porches are determined to be in (G/A/F/P)_____ condition. There is (y/n)_____ evidence of structural problems.

◊ The landing platform(s), which is the area in front of doors, does (y/n)_____ have a large enough space to safely open the door while someone is standing there. This is required to help prevent someone from being knocked down the steps when the door is opened.

◊ There are (y/n)_____ handrails for <u>all</u> stairs that are more than two steps in height. The handrails are (y/n)_____ loose and/or decayed. Handrails **must** be installed when not noted and maintained periodically to prevent any tripping hazards.

◊ The steps do (y/n)_____ have uneven and/or damaged sections that require repairs at this time. All steps *must* have an even and uniform height and be properly maintained so that there are no tripping hazards.

◊ There are (y/n)_____ wood stairs with the wood base resting directly on the soil. When wood stairs are noted, the base of the wood should be resting on concrete pads above the soil. This will help prevent rot and termite infestation.

_____
_____
_____
_____

Walks

◊ All accessible walks are determined to be in (G/A/F/P)_____ condition.

◊ There are (y/n)_____ uneven and/or damaged sections in the walks that require repairs at this time to prevent any tripping hazards. There are (y/n)_____ weeds growing in between the walkway sections that need to be removed.

◊ There is (y/n)_____ a sidewalk at the street. When a sidewalk is noted, it is recommended that you check with the local building department to determine whose responsibility it is to repair the sidewalk. In most areas, the homeowner is responsible for repairing and cleaning the sidewalk in front of their home.

_____
_____
_____
_____

Patios and Terraces

◊ There is (y/n)_____ a patio(s) associated with the subject property.

◊ The patio(s) is determined to be in (G/A/F/P)_____ condition. There are (y/n)_____ uneven and/or damaged sections in the patio(s) that require repairs at this time to prevent any tripping hazards. There are (y/n)_____ weeds growing in between the patio sections that need to be removed.

◊ The joints of the patio are (y/n)_____ properly caulked. If the patio touches the side of the foundation, then it should be well caulked and sloped away to prevent water from draining towards the house.

◊ In most areas building permits and approvals are needed to build patios. Determine if they have been obtained for any patio(s) at the site.

_____

_____

_____

_____

Decks

◊ There is (y/n)_____ a deck associated with the subject property. Decks <u>always</u> require building department approvals because of the safety concern if they are improperly built. Determine from the local municipality if all valid permits and approvals have been obtained for any deck construction.

◊ There is (y/n)_____ evidence of rotted and/or damaged sections of wood that need to be replaced.

◊ The deck railings are (y/n)_____ sturdy. The deck railings are (y/n)_____ properly spaced for safety. The deck perimeter railings *must* be spaced so that a maximum gap of 4 inches exists between them. This is to help prevent small children or dogs from falling through the openings.

◊ The main beam is (y/n)_____ lag bolted to the side of the house. There *must* be lag bolts in the main beam, called the *"header beam,"* where the deck is attached to the side of the house. Lag bolts are a **far superior** way to support the deck, as opposed to just using nails.

◊ The floor joists of the deck do (y/n)_____ have steel support hangers. The floor joists of the deck *must* have steel support hangers to give them additional support, as opposed to just nailing them.

◊ The deck support posts and girders do (y/n)_____ have steel support brackets. All deck support posts and girders *must* have steel support brackets at the base **and** at the top for support. The base of the support posts are (y/n)_____ resting on concrete pads. The base of the posts should be resting on a concrete pads with steel brackets to keep the wood from being in contact with the soil.

Walls and Fences

◊ There is (y/n)_____ a wall(s) and (y/n)_____ a fence(s) associated with the subject property.

◊ Retaining walls are used to support the soil in areas that are dug into the earth, such as driveways or yards. The retaining wall(s) noted is made of _____. The wall(s) is in (G/A/F/P)_____ condition. The retaining wall(s) is (y/n)_____ leaning at this time. Any leaning conditions indicate that repairs **must** be made to prevent the wall from moving any further or collapsing.

◊ The retaining wall(s) does (y/n)_____ appear to have adequate weep holes at the base. The purpose of *"weep holes"* is to relieve the pressure by allowing any water that builds up behind the wall to drain safely away.

◊ The fence(s) is made of _____. The fence(s) is in (G/A/F/P)_____ condition. It is *highly* recommended that you check with town hall to determine if the fence is located within the subject property line. Often the homeowner or a neighbor will have a fence installed and the contractor will just guess where the property line is. This will lead to a property line encroachment.

Drainage and Grading

◊ The soil grading next to the foundation is (y/n)_____ properly sloped away from the house. The soil must slope away from the structure to help prevent the rainwater from building up next to the foundation. In most cases, the soil only needs to slope about 1/2 inch for every foot away from the house to properly drain the water.

◊ The bushes, shrubs and/or trees are (y/n)_____ properly pruned away from the side of the house. The bushes, shrubs and trees must be pruned away from the side of the house to allow enough sunlight and air next to the foundation to help prevent rot and wood destroying insect problems.

_____
_____
_____
_____

Driveways

◊ There is (y/n)_____ a driveway on the site. Driveways that do not have a finished surface and are made of gravel and dirt are not recommended. Often they have holes which are a tripping hazard which need to be filled. Unfinished driveways can lead to people tracking dirt into the house and they cannot be shoveled for snow removal in colder areas

◊ The driveway is made of (dirt/gravel/asphalt/concrete/etc) _____
The driveway does (y/n)_____ have holes, cracked and/or uneven sections that need to be repaired. Asphalt driveways need to be sealed with a driveway sealer every 2 or 3 years to prevent them from drying out and cracking. Concrete driveways need to be patched periodically. Gravel driveways get potholes that need filling.

◊ There is (y/n)_____ a water removal drain at the base of the driveway. These drains need to be checked periodically due to problems with clogging.

_____
_____
_____
_____

Garage

◊ There is a (y/n) or (built-in/attached/detached)_____ garage on the premises. The garage has a capacity for (#)_____ cars. The benefit of having an attached or built-in garage is that you can park the car and enter the house without worrying about the weather conditions. A detached garage is safer in the event that a car is left running by mistake, or if there is a garage fire.

◊ The garage was in overall (G/A/F/P)_____ condition. The garage doors did (y/n)_____ come down with excessive force when checked. The door springs will need to be adjusted and lubricated periodically. Doors that come down with excessive force can crush a child if they are caught underneath.

◊ The electric door openers did (y/n)_____ operate properly. The automatic reverse function was tested and did (y/n)_____ operate properly. Electric door openers *must* have an *"automatic reverse"* function that is working properly. This is a setting in the opener that will stop or reverse the direction of the doors if a person or a car gets caught underneath. It must be checked periodically for safety.

◊ All visible ducts and water pipes in the garage were (y/n)_____ insulated. All ducts and water pipes in the garage need to be insulated to prevent them from freezing and for maximum energy efficiency.

◊ There are (y/n)_____ excessive gas and/or oil drippings on the garage floor. Any oil or gas drippings *must* be cleaned to help prevent any fires.

◊ The garage walls and ceilings do (y/n)_____ appear to have an adequate fire resistant covering. The garage walls and ceilings *must* be covered with fireproof sheetrock to help prevent the spread of any fires. Masonry walls and ceilings are an acceptable fireproof covering.

◊ The garage does (y/n)_____ have a fire resistant entry door leading to the house. The door does (y/n)_____ have a properly operating self-closing device. If the garage is attached or built-in there *must* be a fireproof entry door leading to the house. Also, this door *must* have a self-closing device to prevent the door from being left open so car exhaust fumes or fires cannot spread easily into the house.

_____
_____
_____
_____

Other Exterior Structures

◊ There is (y/n)_____ an exterior shed or other structure on the property.

_____
_____
_____

Swimming Pools

◊ There is an (y/n) or (above-ground/in-ground)_____ swimming pool on the property. **All** swimming pools require local town permits and approvals which need to be verified.

◊ There is (y/n)_____ a fence surrounding the pool. **All** swimming pools need to have fences surrounding them to prevent any children from falling into the water and drowning. Special homeowners insurance is needed with swimming pools.

◊ There are (y/n)_____ tripping hazards noted around the pool area which need to be repaired.

◊ The pool walls did (y/n)_____ appear to have evidence of leaks, cracks and/or bulging sections that need to be evaluated by a reputable pool contractor.

_____
_____
_____
_____

Wood Destroying Insects

◊ There is (y/n)_____ evidence of wood destroying insect damage in the visible and accessible areas of the subject property. If there was any aspect of performing a home inspection that the inspector is believed by some clients to have X-ray vision, then this one takes the prize! If *Superman* really existed, he would make a fortune as a termite inspector! Many home inspectors get complaints from former clients because they did not notice termites that were behind the finished and covered walls and/or floors. **THE INSPECTOR CANNOT BE HELD RESPONSIBLE FOR AREAS AND ITEMS THAT ARE NOT ACCESSIBLE OR NOT VISIBLE!!**

◊ There are many different types of wood destroying insects, including 70 species of termites throughout the world. The wood destroying insects that generally concerns people the most are: Subterranean Termites, Dry Wood Termites, Damp Wood Termites, Powder Post Beetles, Carpenter Ants, and Carpenter Bees.

- o ***Termites*** eat the wood and turn it into food. They have one celled organisms in their digestive tracts which converts the cellulose of wood back into sugar which they can digest. In forests termites are beneficial in the fact that they help to decompose fallen trees and stumps and return the wood substances to the soil to be used again by other trees. Termite damaged wood will have channels in it and there will not be any sawdust around.

- o ***Powder Post Beetle*** larvae eat the wood and lay their eggs in it. They <u>cannot</u> convert the cellulose in the wood to sugar and therefore, must get their nourishment from the starch and sugar which the tree had stored in the wood cells. To these insects the cellulose in the wood has no food value and is thus ejected from their bodies as wood powder or *"frass."* They derive nourishment from the starch and sugar in the wood.

- ***Powder Post Beetle*** damaged wood will crumble like sawdust when you probe it. A common indication of these insects is the existence of tiny holes in the wood. If only a single generation of this beetle larvae has fed within some wood, it is usually still structurally sound. But the feeding of generation after generation is what reduces the interior of the wood to a mass of powder. Before the female will attach her eggs to a piece of wood, she first actually tastes the wood to be sure it contains enough sugar and starch to nourish her offspring. If she is prevented from doing this due to any covering on the wood, such as paint, varnish, stain, etc, she will not deposit her eggs in the wood and it will not be reinfested with another generation of Powder Post Beetle larvae. That is why there should not be any untreated wood around the house.

- ***Carpenter Ants*** and ***Carpenter Bees*** merely excavate the wood to make nests. The damage they cause will leave sawdust outside the wood channels.

◊ An important fact to remember when getting a corrective wood destroying insect treatment on a house is that it is recommended that <u>all</u> of the damaged wood be removed and replaced. This will ensure that any damaged areas of wood are re-supported with good, solid lumber. Another reason for this is that there is *no way* to tell down the road if the wood had gotten the termite damage before or after the corrective treatment was performed.

◊ They say there are 2 kinds of houses: **Houses that <u>have</u> termites; and Houses that <u>will have</u> termites.** That's a fact. All houses will get termite damage to some extent eventually. Sometimes builders will install a termite shield along the top of the foundation wall. A *"termite shield"* is a small metal guard or molding placed at the top of the foundation walls. It is similar in purpose to installing a cap plate at the top of concrete block walls. However, these shields **do not** prevent termites. The only benefit from them is that they *might* deter termites or make it a little more difficult for them to reach the wood.

◊ There are many ways to help prevent wood destroying insect and rot damage:

- Use pressure treated lumber whenever replacing or constructing anything on the site or in the house. Pressure treated lumber has a greenish color to it. It is rot and termite resistant for up to 40 years. The most common type of pressure treated wood is called *CCA 40*. There is also a *CCA 60* pressure treated lumber that has a higher pressurization and life expectancy than CCA 40 does.

- The way the pressure treated process works is they will take the lumber and place it in large vats of chemicals where it will sit until the chemicals are absorbed sufficiently into the wood. The chemicals they use are *copper, chromate and arsenic*. The arsenic deters any wood destroying insects. The type of wood that is used for pressure treated lumber is *Southern Yellow Pine*. The reason for this is that it is the best lumber to use for the chemical process performed.

- Another way to help prevent wood destroying insect and rot damage is to keep all wood siding and trim work about 8 inches above the soil to make it more difficult for termites to get to their food source. Keep all bushes and shrubs pruned and keep the soil and drainage leaders sloped away from the foundation. This will help prevent any dark and moist areas that attracts termites.

- Get a *preventive* wood destroying insect treatment before finding any damage, as opposed to a *corrective* treatment after the damage is found. Preventive treatments are usually about **half** the cost of a corrective treatment because they only treat around the exterior perimeter of the house.

◊ It is *highly* recommended that you get a corrective treatment if the inspector finds damage and a preventive treatment if the inspector does not find any damage. Since all houses do get some form of wood destroying insect damage over time, you might as well eliminate the problem ahead of time when it is less expensive to do so. It is also easier to sell a house with a preventive treatment before any damage is found, as opposed to a corrective treatment done after there was damage found.

◊ There are certain houses that many Pest Control Operators, *(PCO)*, **will not** treat, or else there will be only a limited number of them that will treat the house. Some of these houses that can be difficult to treat are:

- Houses with on-site well water systems. The PCO has to worry about contaminating the well water supply. If the well is less than 100 feet from the house your chances of finding a PCO to treat diminish even further.

- Houses that have brick foundation walls. The PCO has to worry about contaminating the house by seepage through the brick walls.

- Houses that have air ducts embedded in the lower level cement floor for the heating or air-conditioning systems. The PCO has to worry about contaminating the air ducts.

- If the inspection is being conducted on a condominium then the By-Laws or Prospectus of the Condo/Owner's Association may have requirements that can in some way restrict wood destroying insect treatments.

Additional Comments

Safety Concerns #2

Safety Concerns

◊ Items such as tripping hazards in the steps, walks and patios, loose and missing handrails, proper deck construction and guardrails, leaning retaining walls, and loose electrical grounding cables, can sometimes seem like minor items to repair. However, these are things that can cause someone to get **seriously** hurt if they are not repaired immediately and properly.

◊ An uneven section in a walkway might not seem like much but what happens if the person that falls hits their head. A leaning retaining wall will crush a child if it falls on top of them! A missing or loose handrail could cause someone to fall down the steps. **The point we want to make very clear is don't tank chances with safety items!!** You could end up costing someone their *LIFE!!* Repair all hazardous conditions immediately!

Additional Comments

Home Inspection Conclusion #2

Conclusion

◊ Many times the Realtor, the seller or some third party to a transaction that is involved with the deal, will tell the inspection client that something, such as, the roof, the heating system, etc., was just recently replaced. Or many times they will say that all the building permits and approvals have been obtained for an addition, deck, or some other aspect that was a change to the original construction of the building or site. If this is the case with the subject property, then it is *highly* recommended that you obtain all receipts and documentation for the work performed and that you check with the local building department to make sure that this information is accurate!! Whenever you upgrade the roof, heating system, air-conditioning system, electrical system, etc.; put an addition on a house; add a deck; install a swimming pool; or make any changes to a house or a site from the original construction, you have to file the necessary permits with the local municipality. The reason for this is that the local building department inspectors have to check the work to make sure it meets all the necessary building codes in that town.

◊ **Do not** just take it for granted that the permits and approvals have been obtained for any work performed!! Many people will upgrade from the original construction without filing for permits. They might do the work themselves or else they hire a contractor who does not know what he is doing and he will not file any permits for the work performed. You should go down to town hall *personally* to check **all** records pertaining to the subject property!! This will enable you to verify all information in the real estate listing and what has been told to you about the subject property. If you send a Realtor or another third party to town hall to check the records, and they miss something, it is **YOU** that is going to have to deal with the problem later!! This will end up costing you time and money. So you should go and check it yourself, as opposed to just sending someone else to do it for you. At town hall the records will show the amount of taxes on the house, if there are any building violations, any easements, encroachments or problems with the title and deed of the property, and a lot more. All of this information is very valuable to you and many people do not even realize how much information is open for the public to view at their local town hall records department.

◊ The decision to buy or not to buy, and what repairs are done to the subject property is totally up to you. A home inspector can inform you of the current condition of the accessible and visible areas of the subject property only! A home inspector *is not* an appraiser determining **market value** of the subject property. A home inspector only determines the **condition** of the subject property. There is a big difference between the two. The point is that only a very well trained and qualified Real Estate Appraiser can determine market value - not a home inspector.

◊ Any aspects of concern brought forth in the on-site inspection or in the written report *must* be checked out by a reputable, licensed contractor if any doubts exist. You are encouraged to call a contractor and obtain written estimates, on their own, for any areas of concern or repairs needed.

This written report is not assignable to any third parties in any way, shape or form.
No part of this work shall be reproduced, stored in a retrieval system, or transmitted by any means, electronic, mechanical, photocopying, recording or otherwise.
This report format is © Copyright 1992-2004 www.nemmar.com . All rights are reserved.

More Nemmar Products

Email info@nemmar.com for prices

Energy Saving Home Improvements From A to Z ™

Don't let your dream house be a money pit in disguise! Our **5-star rated** book that teaches you how to **save** thousands of dollars **and** help the environment by making minor improvements to your home. You'll learn how to **lower your utility bills by 50%,** live more comfortably, and help the environment. Includes many photographs with detailed descriptions.

Home Inspection Business From A to Z ™

The REAL FACTS the other books don't tell you! Our **number one** selling home inspection book. This is **definitely** the best home inspection book on the market and has been called the "Bible" of the inspection industry. *Every* aspect of home inspections is covered with precise steps to follow. Includes many photographs with detailed descriptions.

Real Estate Appraisal From A to Z ™

The REAL FACTS the other books don't tell you! Our **number one** selling appraisal book. This is **definitely** the best real estate appraisal book on the market. *Every* aspect of real estate appraising is covered with precise steps to follow. Includes sample professional appraisal reports and many photographs with detailed descriptions.

DVD's - Home Inspection From A to Z ™

Our **5 star rated** DVD's have two hours of video plus you get the 80 page *HIB **DVD** Companion Guidebook!*
    OPERATING SYSTEMS DVD topics including: heating systems, air-conditioning, water heaters, plumbing, well water system, septic system, electrical system, gas service, and auxiliary systems. Health Concerns topics including: asbestos insulation, radon gas, and water testing.
    INTERIOR and EXTERIOR DVD topics including: roof, chimneys, siding, eaves, gutters, drainage and grading, windows, walkways, entrances and porches, driveways, walls and fences, patios and terraces, decks, swimming pools, exterior structures, wood destroying insects, garage, kitchen, bathrooms, floors and stairs, walls and ceilings, windows and doors, fireplaces, attics, ventilation, insulation, basement/lower level, and water penetration.

Home Buyer's Survival Kit ™

Don't buy, sell, or renovate your home without this! Includes: Four of our **top selling** books – *Real Estate Home Inspection Checklist From A to Z, Energy Saving Home Improvements From A to Z, Home Inspection Business From A to Z,* and *Real Estate Appraisal From A to Z.* Plus, you get both of our *Home Inspection From A to Z* – **DVD's**. As an added bonus you also get the 80 page *HIB **DVD** Companion Guidebook.*

Narrative Report Generator and On-Site Checklist

The report generator and checklist the others don't have! CD-Rom with the *best* Narrative Report Generator and On-Site Checklist on the market! These will enable you to *easily* do 30 page narrative, professional home inspection reports to send to your clients. These will assist you at the inspection site to be sure that you properly evaluate the subject property. Designed to walk you through the entire inspection process with very detailed instructions on how to properly evaluate the condition and status of **all** aspects of a home in a fool-proof, step-by-step system and create professional, narrative reports.

Just some of our books, CD's, DVD's and much more!
Email **info@nemmar.com** for prices.
Visit us at **www.nemmar.com**

Everything You Need To Know About Real Estate From Asbestos to Zoning ™

Real Estate Home Inspection Checklist From A to Z ™

Energy Saving Home Improvements From A to Z ™

Home Inspection Business From A to Z ™

Real Estate Appraisal From A to Z ™

Real Estate From A to Z ™

Nemmar Real Estate Training
info@nemmar.com
www.nemmar.com

Index

100 amp, 50, 130
110 volt, 49, 129
200 amp, 49, 129
220 volt, 49, 129
60 amp, 49, 129
A to Z, 1, 2, 3, 4, 9, 13, 14, 15, 19
A to Z Home Inspector, 14, 15, 19
abnormal, 22, 53, 55, 57, 61, 62, 63, 66, 133, 135, 137, 141, 142, 143, 146
accidents, 15
acidity, 46, 126
additions, 30, 31, 110, 111
agent, 15
aging, 64, 73, 74, 75, 144, 153, 154, 155
agreement, 15
air ducts, 36, 40, 84, 116, 120, 164
air fill valve, 46, 126
air filters, 36, 40, 116, 120
air handler, 40, 120
air intake, 37, 41, 117, 121
air pocket, 38, 118
air sample, 68, 69, 148, 149
air systems, 36, 116
air temperature, 18, 28, 40, 108, 120
air-conditioner, 40, 120
air-conditioning, 15, 16, 17, 18, 21, 28, 30, 31, 37, 40, 41, 68, 84, 88, 108, 110, 111, 117, 120, 121, 148, 164, 168
air-to-water ratio, 46, 126
alarm, 57, 137
aluminum, 68, 148
American Society of Home Inspectors, 16
amperage, 50, 51, 130, 131
amperage rating, 50, 130
analogy, 16, 17
analysis, 16, 18, 19, 64, 70, 144, 150
animals, 74, 76, 154, 156
antenna, 74, 154
appliances, 28, 50, 51, 61, 108, 130, 131, 141
approvals, 29, 31, 46, 48, 50, 53, 55, 56, 78, 82, 88, 109, 111, 126, 128, 130, 133, 135, 136, 158, 162, 168
architect, 16, 17, 19
architecture, 16, 17
arsenic, 83, 163
asbestos, 8, 9, 15, 29, 69, 70, 109, 149, 150, 176
ASHI, 16, 17
asphalt, 80, 160
attic, 21, 64, 66, 67, 68, 75, 144, 146, 147, 148, 155
attic gable, 67, 147
attic vents, 67, 147
attorney, 21
automatic pilot, 15
automatic reverse, 81, 161
awl, 17
backflow preventer, 38, 58, 118, 138
back-smoking, 29, 34, 65, 109, 114, 145
bacteria, 19, 46, 47, 126, 127
bank, 9, 20, 21

basement, 17, 19, 20, 21, 50, 68, 69, 70, 130, 148, 149, 150
bathroom, 18, 31, 50, 62, 67, 111, 130, 142, 147
beam, 66, 146
bedroom, 18
biggest, 3, 9, 13, 14
binoculars, 18
birds and bees, 67, 147
black iron, 35, 56, 115, 136
blower fan, 36, 37, 40, 41, 116, 117, 120, 121
blown fuse, 35, 49, 115, 129
blown-in, 68, 148
boiler, 17, 20, 30, 34, 35, 38, 39, 43, 110, 114, 115, 118, 119, 123
book value, 20
bowing, 66, 73, 146, 153
branch circuit, 49, 50, 129, 130
branch line, 49, 129
brass, 45, 125
brick, 65, 84, 145, 164
brine, 47, 127
broker, 15
BTU, 33, 113
builder, 1, 9, 14, 83, 163
building code, 14, 15, 16, 28, 56, 73, 88, 108, 136, 153, 168
building department, 46, 48, 50, 58, 64, 77, 78, 88, 126, 128, 130, 138, 144, 157, 158, 168
building inspector, 16
bulging, 82, 162
burner, 33, 34, 35, 39, 44, 113, 114, 115, 119, 124
burner flame, 34, 35, 114, 115
bushes, 37, 41, 80, 83, 117, 121, 160, 163
By-Laws, 84, 164
C of O, 30, 110
cabinets, 61, 62, 141, 142
calculations, 16
camera, 17
cancer, 8, 70, 150
canister, 18, 70, 150
canvas, 36, 40, 69, 116, 120, 149
carbon, 15, 18, 33, 34, 35, 36, 113, 114, 115, 116
carbon monoxide, 15, 18, 33, 34, 35, 36, 113, 114, 115, 116
carpenter ants, 82, 83, 162, 163
carpenter bees, 82, 83, 162, 163
carpets, 29, 54, 63, 109, 134, 143
cast iron, 45, 69, 125, 149
caulk, 62, 142
caulking, 74, 75, 78, 154, 155, 158
CCA, 83, 163
cellulose, 82, 162
cement, 57, 65, 74, 84, 137, 145, 154, 164
central air-conditioning, 19, 40, 120
Certificate of Occupancy, 14, 30, 32, 34, 110, 112, 114
certification, 16, 29, 109
cheat, 15

check valve, 58, 138
chemical, 83, 163
child guards, 61, 62, 141, 142
children, 51, 53, 61, 62, 63, 78, 81, 82, 86, 131, 133, 141, 142, 143, 158, 161, 162, 166
chimney, 18, 34, 35, 65, 74, 114, 115, 145, 154
chimney flue, 65, 145
chimney sweep, 65, 145
chromate, 83, 163
circuit breaker, 35, 49, 115, 129
circulator pump, 38, 118
city sewer, 19
city water system, 19
clog, 43, 48, 57, 76, 80, 123, 128, 137, 156, 160
coils, 37, 41, 43, 44, 117, 121, 123, 124
cold line, 42, 122
cold water, 42, 43, 122, 123
collar beams, 66, 146
colored dye, 48, 128
combustible, 33, 113
combustion, 33, 113
commission, 9
competent, 9
compressor, 30, 37, 41, 110, 117, 121
concrete, 37, 41, 53, 55, 57, 77, 78, 80, 83, 117, 121, 133, 135, 137, 157, 158, 160, 163
concrete block, 83, 163
condensate, 40, 120
condensate pump, 40, 120
condensation, 40, 46, 57, 64, 120, 126, 137, 144
condominium, 15, 16, 18, 21, 29, 31, 84, 109, 111, 164
conduit, 49, 129
consistent, 22
consultant, 9, 15
consulting, 15
contract, 20, 28, 34, 108, 114
contraction, 63, 143
copper, 34, 35, 41, 45, 83, 114, 115, 121, 125, 163
corrective termite treatment, 83, 84, 163, 164
corrosion, 34, 45, 49, 50, 56, 114, 125, 129, 130, 136
cosmetic, 14, 73, 153
countertop, 61, 141
course, 19
cover plate, 44, 124
Cover Your Assets, 21
cracked, 36, 49, 53, 54, 55, 62, 64, 80, 116, 129, 133, 134, 135, 142, 144, 160
cracking, 20, 80, 160
crawl space, 17, 18, 19, 55, 68, 135, 148
creosote, 65, 145
crucial, 1
crystal ball, 16
customer feedback, 13
cut corners, 14

Index

CYA, 21
damaged, 29, 34, 64, 66, 73, 74, 75, 76, 77, 78, 82, 83, 109, 114, 144, 146, 153, 154, 155, 156, 157, 158, 162, 163
damaged areas, 29, 64, 73, 74, 75, 76, 83, 109, 144, 153, 154, 155, 156, 163
Damp Wood Termites, 82, 162
damper, 65, 145
dangerous, 15, 28, 108
data, 5, 17
data plate, 17
deceive, 17
deck, 30, 78, 86, 88, 110, 158, 166, 168
decompose, 47, 127
deed, 88, 168
deep wells, 46, 126
degrees, 16, 17, 28, 40, 42, 44, 67, 108, 120, 122, 124, 147
dehumidifier, 57, 137
depreciate, 20
diagonal bracing, 54, 55, 134, 135
die, 48, 70, 128, 150
differential, 46, 126
dirt floor, 55, 135
disadvantage, 43, 123
discharge, 33, 38, 39, 42, 47, 58, 67, 70, 113, 118, 119, 122, 127, 138, 147, 150
disclosure, 20
disconnect switch, 37, 41, 117, 121
disease, 8
dishonest, 17
distressed sale, 14
documentation, 31, 48, 88, 111, 128, 168
do-it-yourself, 15, 28, 50, 58, 108, 130, 138
dollar, 6, 7, 9, 14, 19
doors, 64, 65, 75, 77, 81, 144, 145, 155, 157, 161
downdrafts, 35, 65, 115, 145
downspouts, 57, 76, 137, 156
draft, 29, 34, 35, 109, 114, 115
draft diverter hood, 35, 115
draft regulator, 34, 114
drain, 38, 39, 40, 42, 57, 62, 76, 79, 80, 118, 119, 120, 122, 137, 142, 156, 159, 160
drain valve, 38, 39, 42, 118, 119, 122
drainage, 20, 31, 38, 40, 45, 48, 57, 58, 62, 76, 83, 111, 118, 120, 125, 128, 137, 138, 142, 156, 163
drainage line, 40, 48, 57, 58, 76, 120, 128, 137, 138, 156
drainage pipe, 48, 58, 128, 138
drill, 46, 70, 126, 150
drip loop, 49, 129
driveway, 79, 80, 159, 160
dry wells, 76, 156
Dry Wood Termites, 82, 162
duct, 36, 40, 68, 81, 84, 116, 120, 148, 161, 164
dye, 18, 19, 48, 128
dye test, 19, 48, 128
E and O insurance, 20
easement, 88, 168
economic, 4
economy, 14
educated, 9, 14
efficient, 34, 42, 68, 114, 122, 148
efficient operation, 34, 114

efflorescence, 57, 137
electric baseboard radiators, 35, 115
electric heating systems, 35, 115
electrical current, 50, 130
electrical ground, 45, 50, 86, 125, 130, 166
electrical grounding cable, 86, 166
electrical hazards, 15, 50, 130
electrical lines, 49, 129
electrical meter, 49, 129
electrical overload, 31, 111
electrical panel, 18, 49, 50, 129, 130
electrical service, 49, 129
electrical service lines, 49, 129
electrical system, 45, 49, 50, 88, 125, 129, 130, 168
electrical wire, 8, 50, 130
electricity, 20, 33, 42, 49, 50, 64, 113, 122, 129, 130, 144
electrified, 49, 129
electrocuted, 49, 51, 129, 131
emergency, 33, 34, 35, 37, 41, 45, 46, 49, 50, 56, 64, 113, 114, 115, 117, 121, 125, 126, 129, 130, 136, 144
emotional, 14
encroachment, 79, 88, 159, 168
energy efficient, 9, 33, 36, 40, 41, 42, 43, 65, 68, 75, 81, 113, 116, 120, 121, 122, 123, 145, 148, 155, 161
engineer, 16, 17, 19
engineering, 3, 16, 17
entrances, 77, 157
environmental, 69, 70, 149, 150
Environmental Protection Agency, 8, 68, 69, 70, 148, 149, 150
EPA, 8, 68, 69, 70, 148, 149, 150
Errors and Omissions, 20
evaporator, 40, 120
evaporator coil, 40, 120
excessive, 22, 35, 36, 38, 39, 40, 42, 45, 46, 49, 50, 53, 54, 55, 56, 57, 65, 66, 67, 76, 81, 115, 116, 118, 119, 120, 122, 125, 126, 129, 130, 133, 134, 135, 136, 137, 145, 146, 147, 156, 161
exhaust, 81, 161
expand, 20, 38, 50, 118, 130
expansion, 38, 63, 118, 143
expansion tank, 38, 118
expensive, 55, 64, 73, 84, 135, 144, 153, 164
expert, 1, 2, 9, 17
expertise, 15
explode, 8, 38, 39, 42, 56, 118, 119, 122, 136
exposed wire, 49, 50, 129, 130
extension cord, 51, 131
extensive, 22, 48, 63, 128, 143
factory, 42, 44, 122, 124
fahrenheit, 40, 42, 44, 67, 120, 122, 124, 147
fail, 48, 128
failure, 37, 41, 42, 47, 48, 117, 121, 122, 127, 128
false, 50, 130
family, 3, 15, 18
fan, 36, 40, 67, 70, 116, 120, 147, 150
faucet, 38, 42, 43, 45, 46, 48, 61, 62, 118, 122, 123, 125, 126, 128, 141, 142
fence, 79, 82, 159, 162
field notes, 17

filter, 32, 34, 40, 47, 61, 112, 114, 120, 127, 141
finished areas, 16
finished covering, 63, 143
fire, 33, 51, 57, 64, 65, 81, 113, 131, 137, 144, 145, 161
fire resistant, 81, 161
firebox, 17, 34, 65, 114, 145
firematic shutoff valve, 34, 114
fireplace, 29, 65, 109, 145
fireproof, 33, 69, 81, 113, 149, 161
flashlight, 17
flood, 58, 138
flood hazard, 58, 138
flood hazard insurance, 58, 138
flood hazard zone, 58, 138
flood maps, 58, 138
floor covering, 19, 61, 62, 141, 142
floor joist, 53, 54, 55, 66, 68, 78, 133, 134, 135, 146, 148, 158
fluctuate, 70, 150
flue, 33, 34, 35, 65, 74, 113, 114, 115, 145, 154
flue lining, 65, 145
flue pipe, 33, 34, 35, 113, 114, 115
flue stack, 74, 154
foam, 29, 109
foil, 68, 148
forced hot water heating system, 38, 118
foreclosure, 14, 20
foundation, 29, 37, 41, 53, 55, 56, 57, 58, 70, 76, 78, 80, 83, 84, 109, 117, 121, 133, 135, 136, 137, 138, 150, 156, 158, 160, 163, 164
foundation wall, 53, 57, 83, 84, 133, 137, 163, 164
freeze, 45, 125
freezing, 81, 161
Freon, 40, 41, 120, 121
fuel, 33, 34, 35, 113, 114, 115
fumigated, 63, 143
furnace, 17, 20, 30, 34, 35, 36, 110, 114, 115, 116
furniture, 29, 53, 109, 133
fuse, 49, 129
gallon, 42, 46, 122, 126
gallons per minute, 46, 126
galvanized iron, 45, 125
garage, 19, 50, 64, 81, 130, 144, 161
gas, 8, 15, 18, 20, 33, 35, 40, 42, 44, 56, 69, 70, 81, 113, 115, 120, 122, 124, 136, 149, 150, 161
gas detector, 18
gas feed lines, 35, 56, 115, 136
gas fired heating systems, 35, 115
gas leaks, 15, 56, 136
gas lines, 35, 56, 115, 136
gas meter, 8, 56, 136
gas service, 56, 136
GFCI, 18, 50, 58, 61, 62, 130, 138, 141, 142
girders, 78, 158
glass, 17, 39, 64, 65, 119, 144, 145
gloves, 17
government, 58, 138
GPM, 46, 126
grading of the soil, 57, 137
Ground Fault Circuit Interrupter, 50, 130
grounding cable, 50, 130
grounding prong, 50, 130
grounding rod, 50, 130

grounding wire, 45, 50, 125, 130
groundwater, 58, 138
groundwater table, 58, 138
grout, 62, 142
growth, 4
guidelines, 22
gutters, 57, 76, 137, 156
handrails, 53, 63, 66, 77, 86, 133, 143, 146, 157, 166
hardwood, 54, 63, 134, 143
hazard, 49, 50, 53, 58, 86, 129, 130, 133, 138, 166
header beam, 78, 158
health, 19, 38, 43, 50, 61, 68, 69, 70, 118, 123, 130, 141, 148, 149, 150
health concern, 68, 69, 70, 148, 149, 150
health hazard, 19, 38, 68, 118, 148
heat exchanger, 17, 35, 36, 38, 39, 115, 116, 118, 119
heat loss, 76, 156
heat pump, 37, 117
heat pump heating systems, 37, 117
heating contractor, 69, 149
heating system, 17, 20, 21, 30, 33, 34, 35, 36, 38, 39, 88, 110, 113, 114, 115, 116, 118, 119, 168
hidden, 29, 63, 109, 143
high pressure line, 41, 121
highly, 3, 14, 19, 28, 34, 46, 48, 51, 58, 61, 64, 66, 68, 69, 76, 79, 84, 88, 108, 114, 126, 128, 131, 138, 141, 144, 146, 148, 149, 156, 159, 164, 168
highly recommend, 19, 28, 34, 46, 48, 51, 58, 61, 64, 66, 68, 69, 76, 79, 84, 88, 108, 114, 126, 128, 131, 138, 141, 144, 146, 148, 149, 156, 159, 164, 168
holding tank, 48, 128
home improvement, 9
home warranty programs, 20
honest, 13, 19
hot water, 38, 42, 43, 44, 61, 62, 118, 122, 123, 124, 141, 142
humidifier, 36, 116
humidity, 57, 66, 67, 137, 146, 147
hurt, 86, 166
immediately, 38, 39, 42, 56, 86, 118, 119, 122, 136, 166
immersed, 43, 44, 123, 124
immersion coil, 43, 123
important, 13, 14, 18, 19, 20, 21, 28, 31, 37, 45, 50, 67, 83, 108, 111, 117, 125, 130, 147, 163
improper wiring, 50, 130
inaccessible, 19, 28, 53, 108, 133
inaccessible areas, 19, 53, 133
income, 4, 15
in-depth, 1
industry, 4, 9
inexpensive, 43, 123
infestation, 55, 77, 135, 157
injured, 15
insect, 19, 31, 82, 83, 111, 162, 163
insulation, 17, 29, 30, 36, 40, 41, 46, 68, 69, 81, 109, 110, 116, 120, 121, 126, 148, 149, 161
insurance, 20, 82, 162
integrity, 13, 17
intelligent, 14
internally, 32, 36, 40, 48, 112, 116, 120, 128
internally inspected, 32, 112
invest, 7, 9
investing, 13
investment, 3, 9, 13, 14
investor, 1, 9, 14
jacuzzi, 62, 142
judgments, 73, 153
jumper cable, 45, 125
kill, 14, 15, 49, 129
kitchen, 50, 61, 130, 141
knowledge, 3, 9, 14, 15, 46, 48, 126, 128
knowledgeable, 1, 17, 21
laboratory, 18, 19, 46, 47, 64, 69, 70, 126, 127, 144, 149, 150
lag bolts, 78, 158
landing platform, 77, 157
lawn, 48, 57, 128, 137
lawn sprinklers, 57, 137
layers of shingles, 30, 73, 110, 153
leaching, 31, 48, 111, 128
leaching field, 31, 48, 111, 128
Lead, 8, 45, 125
lead paint, 64, 144
leaders, 76, 83, 156, 163
leak, 8, 18, 21, 32, 34, 36, 40, 45, 61, 62, 66, 82, 112, 114, 116, 120, 125, 141, 142, 146, 162
leaning, 37, 41, 74, 76, 79, 86, 117, 121, 154, 156, 159, 166
legal, 9
lend, 21
lender, 9, 21
liability, 15, 16, 18, 19
liable, 19
license, 9, 16, 28, 29, 38, 39, 40, 42, 45, 46, 48, 49, 50, 53, 54, 55, 56, 58, 61, 62, 63, 64, 66, 68, 69, 70, 88, 108, 109, 118, 119, 120, 122, 125, 126, 128, 129, 130, 133, 134, 135, 136, 138, 141, 142, 143, 144, 146, 148, 149, 150, 168
licensed appraiser, 9
licensed contractor, 16, 28, 38, 39, 40, 42, 48, 53, 54, 55, 56, 58, 61, 62, 63, 69, 88, 108, 118, 119, 120, 122, 128, 133, 134, 135, 136, 138, 141, 142, 143, 149, 168
licensed electrician, 49, 50, 129, 130
licensed EPA contractor, 29, 109
licensed heating contractor, 38, 39, 118, 119
licensed laboratory, 68, 148
licensed septic contractor, 48, 128
life expectancy, 16, 17, 28, 33, 36, 37, 40, 41, 42, 46, 48, 73, 74, 83, 108, 113, 116, 117, 120, 121, 122, 126, 128, 153, 154, 163
light fixture, 63, 143
light switch, 63, 143
limitations, 20
limited, 15, 18, 20, 21, 28, 50, 57, 61, 62, 64, 84, 108, 130, 137, 141, 142, 144, 164
lining, 65, 74, 145, 154
Liquid Petroleum Gas, 56, 136
listing, 14, 15
listing price, 14
load bearing wall, 74, 154
local codes, 70, 150
locks, 64, 144
low pressure line, 41, 121
LPG, 56, 136
lumber, 83, 163
lung cancer, 8, 69, 70, 149, 150
lungs, 69, 149
magic wand, 16
magnet, 18
main disconnect, 50, 130
main electrical panel, 49, 50, 129, 130
main girder, 53, 55, 66, 133, 135, 146
main panel, 49, 50, 129, 130
main shutoff, 45, 125
main water line, 45, 125
major expense, 56, 136
map, 17
market value, 14, 88, 168
masonry, 81, 161
math, 17
maximum, 28, 36, 40, 42, 49, 70, 73, 78, 81, 108, 116, 120, 122, 129, 150, 153, 158, 161
measure, 18, 63, 70, 143, 150
measurement, 16, 18
mineral, 19, 46, 57, 126, 137
mineral salts, 57, 137
minimum, 14, 33, 46, 50, 113, 126, 130
minor, 18, 57, 63, 70, 86, 137, 143, 150, 166
mirror, 17
mitigated, 70, 150
mitigation, 70, 150
moisture, 55, 64, 67, 68, 75, 135, 144, 147, 148, 155
moisture problems, 55, 67, 68, 135, 147, 148
mold, 73, 153
money, 13, 14, 20, 48, 88, 128, 168
monthly assessment, 18
mortar, 65, 74, 145, 154
mortar joints, 65, 74, 145, 154
mortgage, 9, 21
municipal, 31, 45, 111, 125
municipality, 16, 34, 53, 55, 78, 88, 114, 133, 135, 158, 168
must be, 14, 17, 19, 20, 21, 33, 34, 36, 38, 39, 41, 42, 45, 46, 49, 50, 53, 54, 55, 56, 63, 64, 65, 68, 70, 74, 77, 78, 79, 80, 81, 88, 113, 114, 116, 118, 119, 121, 122, 125, 126, 129, 130, 133, 134, 135, 136, 143, 144, 145, 148, 150, 154, 157, 158, 159, 160, 161, 168
nails, 78, 158
National Electric Code, 8, 50, 130
natural gas, 8
NEC, 50, 51, 130, 131
necessary, 34, 35, 42, 56, 88, 114, 115, 122, 136, 168
negative, 13
neighbor, 70, 79, 150, 159
neutral, 50, 130
neutral wire, 50, 130
nondestructive, 15, 18
obstructions, 37, 41, 117, 121
odor, 48, 128
oil, 18, 20, 30, 33, 34, 36, 38, 40, 42, 44, 48, 69, 81, 110, 113, 114, 116, 118, 120, 122, 124, 128, 149, 161
oil feed, 34, 114
oil filters, 34, 36, 40, 114, 116, 120
oil fired, 33, 113
oil supply, 34, 114

Index

oil tank, 18, 30, 34, 110, 114
operating properly, 16, 28, 37, 38, 39, 46, 64, 65, 108, 117, 118, 119, 126, 144, 145
operating systems, 16, 17, 18, 21, 28, 30, 33, 108, 110, 113
outlet, 8, 18, 20, 31, 50, 51, 58, 61, 62, 111, 130, 131, 138, 141, 142
overheat, 51, 131
overpriced, 14
paint, 64, 75, 76, 83, 144, 155, 156, 163
panel cover, 17
patios, 78, 86, 158, 166
PCO, 84, 164
PE, 17
perimeter, 78, 83, 158, 163
permit, 14, 29, 30, 32, 34, 46, 48, 50, 53, 55, 56, 78, 82, 88, 109, 110, 112, 114, 126, 128, 130, 133, 135, 136, 158, 162, 168
personal items, 53, 133
Pest Control Operator, 84, 164
pets, 63, 143
physics, 17
pico curies, 70, 150
pilot light, 35, 115
pipe joints, 38, 39, 118, 119
plaster, 63, 143
plastic, 51, 55, 131, 135
plenum, 36, 40, 116, 120
plot plan, 31, 32, 46, 48, 111, 112, 126, 128
plumber, 69, 149
plumbing system, 31, 111
plywood, 54, 66, 134, 146
poisoning, 8
pond, 21
pool, 32, 82, 112, 162
positive, 13, 16
potential problems, 15, 16
pounds per square inch, 38, 39, 46, 118, 119, 126
powder post beetle larvae, 82, 83, 162, 163
powder post beetles, 82, 162
precaution, 50, 63, 130, 143
preinspection questions, 29, 109
premature, 37, 41, 42, 47, 117, 121, 122, 127
pre-purchase, 14
pre-sale, 14
pressure gauge, 38, 39, 46, 118, 119, 126
pressure reducing valve, 45, 125
pressure relief valve, 38, 39, 42, 46, 118, 119, 122, 126
pressure treated lumber, 83, 163
prevent moisture, 53, 57, 133, 137
price quote, 17, 18, 19
probe, 17, 83, 163
professional engineer, 16, 17
profit, 7, 9, 13, 14
property line, 79, 159
prospectus, 84, 164
prune, 37, 41, 49, 80, 83, 117, 121, 129, 160, 163
psi, 38, 39, 42, 46, 118, 119, 122, 126
pumped clean, 48, 128
pumped out, 32, 48, 58, 112, 128, 138
purchase, 6, 9, 13, 14, 20
PVC, 45, 125
RA, 17

radiation, 70, 150
radiator, 33, 35, 38, 39, 113, 115, 118, 119
radon, 8, 15, 18, 19, 46, 69, 70, 126, 149, 150
radon canisters, 18
radon contractor, 70, 150
radon gas, 18, 19, 70, 150
radon lab, 18, 70, 150
radon level, 70, 150
rafters, 66, 68, 146, 148
rainwater, 49, 57, 76, 80, 129, 137, 156, 160
rare, 16
rating, 1, 46, 50, 126, 130
real estate business, 13, 15
real estate listing, 21, 88, 168
real facts, 9, 13
recession, 4, 14
recommend, 14, 19, 28, 42, 44, 50, 51, 53, 56, 57, 61, 62, 63, 64, 66, 67, 69, 70, 73, 77, 80, 83, 108, 122, 124, 130, 131, 133, 136, 137, 141, 142, 143, 144, 146, 147, 149, 150, 153, 157, 160, 163
recording, 9, 88, 168
records, 28, 31, 48, 88, 108, 111, 128, 168
recovery rate, 44, 124
registered architect, 16, 17
registers, 33, 40, 113, 120
regulate, 44, 124
relocated, 19
relocation, 14
renovate, 30, 31, 110, 111
rental, 14
rental property, 14
repairman, 15
repointed, 65, 145
reputable, 28, 34, 46, 56, 63, 64, 65, 82, 88, 108, 114, 126, 136, 143, 144, 145, 162, 168
required, 9, 16, 33, 34, 35, 37, 38, 40, 41, 44, 54, 55, 58, 61, 62, 64, 68, 69, 75, 77, 113, 114, 115, 117, 118, 120, 121, 124, 134, 135, 138, 141, 142, 144, 148, 149, 155, 157
requirement, 16, 84, 164
residential, 50, 130
retaining walls, 86, 166
reversed, 42, 50, 122, 130
ridge vent, 61, 147
right, 3, 4, 15, 20, 57, 137
rights, 88, 168
rocks, 70, 150
roof, 16, 18, 20, 21, 28, 30, 66, 67, 68, 70, 73, 74, 75, 76, 88, 108, 110, 146, 147, 148, 150, 153, 154, 155, 156, 168
roof rafters, 66, 68, 73, 146, 148, 153
roof ridge beam, 66, 73, 146, 153
roof sheathing, 66, 73, 146, 153
roof shingle, 66, 146
roof vents, 67, 147
roofer, 17
rot, 17, 55, 57, 75, 76, 77, 80, 83, 135, 137, 155, 156, 157, 160, 163
safe, 3, 15, 19, 33, 49, 50, 63, 65, 66, 77, 79, 113, 129, 130, 143, 145, 146, 157, 159
safety, 8, 9, 34, 35, 38, 39, 42, 45, 46, 49, 50, 53, 54, 55, 56, 58, 61, 62, 63,
78, 81, 86, 114, 115, 118, 119, 122, 125, 126, 129, 130, 133, 134, 135, 136, 138, 141, 142, 143, 158, 161, 166
safety hazard, 8, 9, 50, 130
sagging, 53, 54, 55, 63, 133, 134, 135, 143
sales price, 15, 19
salt, 57, 137
school, 17
screwdriver, 17
screws, 18, 33, 113
security, 64, 144
sediment, 38, 39, 42, 118, 119, 122
seepage pits, 48, 128
self-closing device, 81, 161
septic, 17, 18, 19, 29, 30, 31, 32, 47, 48, 109, 110, 111, 112, 127, 128
septic system, 18, 19, 29, 31, 47, 48, 109, 111, 127, 128
septic tank, 31, 32, 47, 48, 111, 112, 127, 128
septic test, 48, 128
serious, 13, 19, 74, 154
service entrance head, 49, 129
service entrance line, 49, 50, 129, 130
setting, 33, 42, 44, 81, 113, 122, 124, 161
settlement, 53, 55, 61, 62, 63, 64, 133, 135, 141, 142, 143, 144
settlement cracks, 53, 55, 61, 62, 63, 64, 133, 135, 141, 142, 143, 144
sewer, 31, 111
sewer system, 31, 111
shaling, 17
shallow wells, 46, 126
sheathing, 66, 146
sheetrock, 29, 33, 63, 64, 81, 109, 113, 143, 144, 161
shingles, 17, 73, 153
shower, 38, 43, 50, 62, 118, 123, 130, 142
shrub, 80, 83, 160, 163
shutoff switch, 33, 34, 35, 46, 113, 114, 115, 126
shutoff valve, 35, 45, 115, 125
sidewalk, 77, 157
siding, 17, 18, 29, 30, 67, 74, 75, 83, 109, 110, 147, 154, 155, 163
significant, 22
single family, 18, 42, 122
sink, 61, 62, 141, 142
site, 9, 13, 16, 17, 18, 19, 21, 22, 28, 31, 34, 56, 57, 78, 80, 83, 84, 88, 108, 111, 114, 136, 137, 158, 160, 163, 164, 168
slate, 17
slate roof, 17
smell, 56, 136
smoke, 64, 65, 144, 145
smoke detector, 64, 144
snow, 21, 80, 160
soffit vent, 67, 75, 147, 155
soil grading, 80, 160
southern yellow pine, 83, 163
span the water meter, 45, 125
species, 82, 162
square footage, 18
stain, 57, 63, 64, 66, 75, 76, 83, 137, 143, 144, 146, 155, 156, 163
staircase, 63, 143
stairway, 63, 66, 143, 146

state, 14, 16, 17, 69, 70, 149, 150
steam, 38, 39, 42, 118, 119, 122
steel, 78, 158
steel brackets, 78, 158
steel support hangers, 78, 158
storage tank, 32, 46, 112, 126
stream, 21
structural, 16, 17, 21, 30, 61, 62, 63, 65, 74, 77, 110, 141, 142, 143, 145, 154, 157
structural problem, 30, 61, 62, 63, 65, 74, 77, 110, 141, 142, 143, 145, 154, 157
structural settlement, 63, 143
subject property, 16, 18, 19, 20, 22, 28, 40, 46, 48, 53, 55, 56, 57, 58, 74, 78, 79, 82, 88, 108, 120, 126, 128, 133, 135, 136, 137, 138, 154, 158, 159, 162, 168
Subterranean termites, 82, 162
suggestions, 13
sump pit, 58, 138
sump pump, 31, 58, 70, 111, 138, 150
sun, 73, 80, 153, 160
Superman, 82, 162
supply ducts, 36, 116
supply lines, 8
support post, 53, 55, 78, 133, 135, 158
surface, 55, 80, 135, 160
survey, 21, 31, 32, 34, 46, 48, 111, 112, 114, 126, 128
swimming pool, 32, 82, 88, 112, 162, 168
switches, 20, 28, 31, 44, 50, 108, 111, 124, 130
tank, 18, 34, 38, 42, 44, 46, 48, 56, 86, 114, 118, 122, 124, 126, 128, 136, 166
tax, 88, 168
temperature, 34, 40, 42, 44, 67, 114, 120, 122, 124, 147
temperature/pressure relief valve, 42, 122
termite, 17, 19, 21, 31, 55, 75, 77, 82, 83, 111, 135, 155, 157, 162, 163
termite damage, 17, 83, 163
termite shield, 83, 163
thermal window, 64, 76, 144, 156
thermometer, 18, 40, 120
thermostat, 15, 28, 33, 40, 42, 44, 67, 108, 113, 120, 122, 124, 147
third party, 20, 21, 88, 168
three prong, 50, 61, 62, 130, 141, 142
tile, 54, 65, 134, 145
title, 58, 88, 138, 168
title insurance, 58, 138

tool, 15, 17, 28, 108
town hall, 28, 30, 31, 56, 58, 79, 88, 108, 110, 111, 136, 138, 159, 168
trees, 37, 41, 49, 73, 80, 82, 117, 121, 129, 153, 160, 162
tripping hazard, 34, 53, 63, 77, 78, 80, 82, 86, 114, 133, 143, 157, 158, 160, 162, 166
tub, 50, 62, 130, 142
tune up, 34, 38, 114, 118
two prong outlets, 50, 130
type of construction, 48, 70, 128, 150
UFFI, 8, 29, 68, 109, 148
underground system, 18
unethical, 14, 17
uneven, 63, 77, 78, 80, 86, 143, 157, 158, 160, 166
up to code, 14
upgrading, 14
upward pitch, 33, 113
Urea Formaldehyde Foam Insulation, 68, 148
urgent, 13
utility, 8, 20, 45, 49, 56, 125, 129, 136
utility company, 8, 45, 49, 56, 125, 129, 136
vacant, 19
vacuum, 57, 64, 69, 137, 144, 149
vacuum seal, 64, 144
value, 1, 14, 20, 82, 88, 162, 168
valve, 38, 39, 42, 45, 56, 118, 119, 122, 125, 136
vapor barrier, 68, 148
varnish, 83, 163
ventilation, 37, 41, 55, 67, 117, 121, 135, 147
vents, 18, 40, 67, 75, 120, 147, 155
very limited, 20, 46, 48, 126, 128
violation, 15, 29, 88, 109, 168
volt, 49, 129
voltage, 18, 49, 50, 129, 130
voltage tester, 49, 129
volume, 32, 112
walks, 69, 77, 86, 149, 157, 166
walkway, 77, 86, 157, 166
wallpaper, 64, 144
warnings, 68, 148
warranty, 20
water analysis, 19, 46, 47, 126, 127
water bottles, 18
water damage, 63, 143
water filter, 47, 61, 127, 141
water freezing, 20
water heater, 20, 42, 43, 122, 123
water leak, 20, 30, 42, 62, 66, 74, 110, 122, 142, 146, 154

water level, 39, 119
water lines, 45, 46, 61, 125, 126, 141
water main, 45, 50, 125, 130
water meter, 45, 125
water meter reading device, 45, 125
water penetration, 31, 57, 74, 75, 111, 137, 154, 155
water pipes, 20, 43, 81, 123, 161
water pressure, 20, 31, 38, 45, 62, 111, 118, 125, 142
water pressure reducing valve, 38, 118
water problem, 17, 20, 57, 76, 137, 156
water sample, 46, 126
water softener, 47, 127
water stains, 57, 63, 66, 137, 143, 146
water storage tank, 46, 126
water supply, 20, 38, 46, 84, 118, 126, 164
water usage, 45, 125
waterlogged, 38, 118
weather, 19, 20, 21, 28, 49, 74, 81, 108, 129, 154, 161
weep holes, 79, 159
weight, 66, 73, 76, 146, 153, 156
well pressure gauge, 46, 126
well pump, 32, 46, 112, 126
well system, 46, 126
well test, 46, 126
well water, 19, 32, 46, 84, 112, 126, 164
well water system, 19, 32, 46, 84, 112, 126, 164
well water test, 46, 126
wells, 17, 18, 30, 46, 110, 126
whole house fan, 67, 147
wind, 74, 154
window frames, 76, 156
window guard, 63, 143
windows, 63, 64, 70, 75, 76, 143, 144, 150, 155, 156
winterized, 20, 32, 112
wire, 45, 50, 51, 125, 130, 131
Wizard of Oz, 16
wood boards, 53, 54, 66, 133, 134, 146
wood destroying insect, 19, 53, 55, 80, 82, 83, 84, 133, 135, 160, 162, 163, 164
worth, 4, 13
written inspection report, 28, 108
written report, 18, 21, 28, 48, 88, 108, 128, 168
X-ray, 16, 70, 82, 150, 162
X-ray vision, 16, 82, 162
zone, 33, 40, 58, 113, 120, 138
zoning, 9, 29, 109, 176

☺ As you can see the last word in this book is "**zoning**" in the index. Like I said, my books cover:

Everything You Need To Know About Real Estate From A̲sbestos to Z̲oning!